Structural Design and Modelling of Flexible Pavement Containing Cold Bituminous Emulsion Mixtures

By

Oluwaseyi Lanre Oke (PhD, Nottingham)

Civil Engineering Department, Ekiti State University, Ado-Ekiti, Ado-Ekiti, Ekiti State, Nigeria
+2348052682689, seyioke@hotmail.com

1

Abstract

This book looks at the effect of cold bituminous emulsion mixtures containing recyclates when used as the road base on the performance of a flexible road pavement in terms of equivalent standard axel loads. The mechanistic-empirical method of flexible pavement design was used and modelling tools used were the KENLAYER, BISAR 3.0 and OLCRACK. The hypothetical examples presented showed that KENLAYER, though time consuming, may be better than BISAR 3.0 for analysing pavements containing CBEMs as road base layers since the materials are non linear in elastic response. When KENLAYER seemed not to be sensitive enough in analysing the pavements with thin surfacing layer thicknesses of 30mm for horizontal strains at the bottom of the surfacing layer, OLCRACK proved in this regard, though the stiffness values calculated in the KENLAYER had to be used. However, in order for OLCRACK to have better application in modelling pavements containing non linear elastic layers, experimental modification reflecting the non linearity of the CBEMs would have to be introduced to the software. OLCRACK is basically good for estimating fatigue lives to failure for linear elastic layers in multi-layer analysis of flexible pavements.

Acknowledgement

Sponsorship by the Ekiti State University, Nigeria and the University of Nottingham, UK, at the instance of which this study was made possible is gratefully acknowledged.

I very much appreciate my supervisors, Dr. T. Parry and Dr. N. H. Thom for being there always. I cannot thank them both enough for their time, advice, guidance, insight, inspiration, concern, and assistance which kept me on track for the entire duration of my study. My sincere thanks are due to Prof. G. D. Airey whose advice and constructive criticisms contributed in no small measure to the success of this study.

The particular help of Dr. D. Day is gratefully acknowledged. Apart from his advice, and constructive criticisms which were helpful, he was instrumental in the support given to this study by Nynas Bitumen, UK and Cliffe Hill Quarry. The support of Longcliffe Quarries, UK is also duly appreciated.

The help of Dr J. Grenfell is deeply appreciated. The assistance of Mr A. Cooper and Mr. K. Maharjan of Cooper Research Technologies, UK are gratefully acknowledged.
Sincere thanks are due to Mr. R. Blakemore, Mr. L. Pont, Mr M. Winfield, Mr J. Watson and their teams, in the Nottingham Transportation Engineering Centre (NTEC) laboratories. I am very thankful for their cooperation and assistance that I enjoyed during my study. Similarly, I gratefully acknowledge the assistance rendered by Mr. M. Roe, helping with numerous sessions on the SEM at the initial stage of this study and Mr A. Batchelor for his kind help, both of the Faculty of Engineering, University of Nottingham.

Oluwaseyi Lanre OKE (PhD, Nottingham)

Table of Contents

List of Figures

List of Tables

1 Overview

The structural design of a flexible road pavement is required in order to estimate its design life. This in essence will ensure that the road pavement serves its purposes structurally and functionally in an economically viable manner within such estimated/predicted design life.

This design task for road pavements is complex (Ebels, 2008; and Twagira, 2010) and it is not as straightforward as obtained in other engineering structures (Thom, 2008). In fact Thom (2008) opined that there is never a unique solution under such circumstance. More importantly, long term characteristics of a pavement have to be borne in mind as much as the day one performance in the process of design. He further suggested that such design should aim to:

- Protect the subgrade
- Guard against deformation in the pavement layers
- Guard against break up of pavement layers
- Protect from environmental attack
- Provide suitable surface
- Ensure maintainability

Researchers have realised that once fatigue and rutting (cumulative deformation) which are the two classical failure modes of pavements have been taken care of in design, then those earlier listed design aims by Thom (2008) are conveniently met (Read and Whiteoak, 2003). Thus, the structural design detailed in this book is mainly focused on fatigue and rutting.

Although researchers have commonly agreed that such design could be done empirically or by using mechanistic approach (Jitareekul, 2009; Ebels, 2008; Twagira, 2010), Huang (2004) in a more elaborate manner stated that there are five methods for executing such design. These are:

- Empirical method with or without a soil strength test
- Limiting shear failure method
- Limiting deflection method
- Regression method based on pavement performance or road test
- Mechanistic – empirical method

The mechanistic-empirical method of the listed five is particularly appealing and has been used for the works described in this book mainly because of the procedure involved. For example, Huang (2004) reported that the mechanistic-empirical method is based on the mechanics of materials that relates an input, such as a wheel load, to an output or pavement response such as strain or stress. The response values are subsequently used to predict distress from laboratory-test and field performance data. This aspect essentially makes this method suitable since many laboratory tests have been conducted in this present work and the results of such are useful in this regard. Huang (2004) believes that dependence on observed values is important because theory alone has not proven sufficient to design pavements realistically. The advantages of the mechanistic methods therefore are the improvement in the reliability of a design, the ability to predict the types of distress, and the feasibility to extrapolate from limited field and laboratory data.

Similarly, following the mechanistic-empirical method described above for the design of pavements containing the five (5) CBEMs (as a road base layer material) allows for a comparative analysis to ascertain the benefit of using RAP CBEMs or otherwise compared to the VACBEM. Although the study here is focused mainly on the road base layer which contains the CBEMs, Ebels (2008) suggested that the responses of interest for a full pavement analysis are:

- Horizontal tensile strain at the bottom of the HMA (surface layer)
- Principal stresses in the CBEM layer (road base)
- Horizontal strain at the bottom of the CBEM layer
- Vertical strain on top of the subgrade.

This book essentially focuses on examining the effect of CBEM type on the relevant responses (stresses and strains) using 16 pavement structures. Table 1 details the pavement structures. The structures have been chosen such that they represent the common practice in the industry. Flexible pavement failure criteria for fatigue and rutting are first discussed before briefly looking at the relevant modelling tools and requirements. A detailed description of non-linear elastic analysis is similarly given in succeeding sections. Since one of the objectives of this book is to check the performance of the CBEMs, the structural analysis started by examining the stress and strain distribution in pavements containing the CBEMs using non-linear modelling tools because the resilient modulus test results suggested this form of behaviour. This is later extended to linear modelling tools for a comparative analysis in order to ascertain the suitable method of analysis for pavements containing CBEMs. Similarly, the maximum horizontal/vertical strains within the CBEM layers and those generated at the bottom of the layers were compared and the more suitable approach for structural analysis was adopted. The effect of the CBEMs on the fatigue life of the surfacing layer and as well the rutting potential of the subgrade is similarly examined and the results are used to produce hypothetical design charts. Although CBEMs both in the early life (EL) and fully cured (FC) condition were studied, only the FC ones have been reported here, since fatigue tests were only conducted on the FC CBEMs.

Table 1: Pavement Layer Thicknesses used for the 16 cases Modelled

Pavement Structure Case	HMA (mm)	Road Base – CBEM (mm)	Sub Base (mm)
1	30	150 (3)*	200
2	50		
3	75		
4	100		
5	30	200 (4)*	200
6	50		
7	75		
8	100		
9	30	250 (5)*	200
10	50		
11	75		
12	100		
13	30	300 (6)*	200
14	50		
15	75		
16	100		

*: Implies number of sub-layer divisions for non-linear analysis

2   Flexible Pavement Failure Criterion

As earlier mentioned, researchers have established that the basic failure mechanisms of flexible pavements are fatigue and rutting related. Therefore, a failure criterion for a flexible pavement must be set in concert with the results obtained for the fatigue and rutting responses of the CBEM materials. The fatigue line equations from ITFT and the parameters obtained from the resilient modulus test are relevant here.

In the case of fatigue failure in asphalt pavements, Thom (2008) observed that a relationship is generally found between tensile strains in asphalts (at the bottom of the asphalt, immediately under the load or; near the surface, just outside the loaded area or; at the surface in the tyre tread contact zone) under traffic load and fatigue tests conducted in the laboratory. In this regard, cracking in asphalt pavements in-service as a result of tensile strains is deemed to be related to failure in fatigue tests under repeated loading. Shift factors which cater for field (in-service) effects must be introduced, i.e. establish a relationship between laboratory and field performance, for a realistic criterion to be set. Oliveira (2006) opined that this is very difficult to do since such depends on the type of test, mode of loading, testing temperature and type of mixture. He observed that laboratory fatigue tests for example use sinusoidal loading and fixed strain or stress during one test, while in practice, the mode of loading is randomly distributed, including rest periods and lateral distribution of loads. He stated further that temperature variations in the asphalt layer and healing effects, due to intermittent loading, also influence the field performance of asphalts.

Oliveira (2006) reported that researchers in the light of the above have suggested various shift factors but with high diversity and no generally accepted one. Although his work focused on grouted macadam, he eventually found a shift factor of 41.4 suitable for fatigue failure criterion for grouted macadam. In a related work to the one described in this thesis, Jilareekul (2009) used the shift factors of 440 (1.1 for lateral wheel load distribution, 20 for rest periods and 20 for crack propagation) and 77 suggested by Brunton (1983) for adjusting laboratory determined fatigue lives (the fatigue lives are multiplied by the shift factors) to suit failure and so-called critical points in the field respectively.

Meanwhile, for rutting, Yusof (2005) and Rahman (2004) among a host of other researchers suggested that by limiting the maximum vertical strain at the top of the sub-grade layer, a failure criterion can be established. In this regard, Brown and Brunton (1986) suggested that:

$$N_{cr} = fr \left(\frac{7.6 \, x \, 10^8}{\varepsilon_z^{3.7}} \right)$$

1

$$N_F = fr \left(\frac{3 \, x \, 10^9}{\varepsilon_z^{3.57}} \right)$$

2

Where:

- N_{cr} = life to critical condition (10mm rut)
- N_F = life to failure (20mm rut)
- Fr = a rut factor to account for the hot mixture type- 1.56 for dense bitumen mixtures
- ε_z = Strain on top of subgrade obtained from structural analysis

However, Thom (1996) cautioned that of these criteria, subgrade strain is less fundamentally based than asphalt tensile strain for fatigue cracking.

These shift factors for fatigue and rutting were derived for hot mixtures, however they have been used in this work since there are no such values for cold mixtures presently.

3   Modelling Tools and Requirements

For the structural design approach chosen in this work i.e. the mechanistic-empirical approach, Brown and Brunton (1986) summarised the design process as follows:

- Specify the loading (vehicular loading),
- Estimate the components (layers) and sizes (thicknesses),
- Consider the materials available,
- Conduct structural analysis using theoretical principles,
- Compare critical stresses, strains or deflections with allowable values (based on the result of laboratory tests) to ascertain whether the design is satisfactory,
- Make adjustments to materials or geometry until satisfactory design is achieved,
- Consider the economic feasibility of the result.

Figure 1 shows the hypothetical flexible road pavement structure that was used for the exercise described in this book though the main focus was on the road base (CBEM) layer. This structure is commonly used in Nigeria and for ease of analysis, a single wheel loading system of 40kN tyre load (80kN being the single standard axle load), 700kPa tyre pressure and tyre contact area radius of 143mm was used for simple symmetry of loading involved (Ebels, 2008). The single wheel loading makes the critical horizontal position to lie directly under the wheel load. More importantly, Twagira (2010) observed that the use of a single wheel also leads to a slightly more conservative estimate, because the stresses under single wheel loading are generally slightly higher than under dual wheel loading.

The results obtained under resilient modulus testing indicated that the CBEMs are stress dependent and thus are non-linear in elastic response. Thus, the work described in this b was analysed bearing in mind that the layer of interest (road base) is non-linear in elastic behaviour. Although some researchers (Jitareekul 2009; and Twagira, 2010) have found linear analytical tools useful for quick analytical design of stresses and strains in such pavements, Ebels (2008) however in a comparative analysis using both linear (BISAR 3.0) and non-linear (KENLAYER) analytical tools on similar materials (cold mixtures) found that it is better to use the appropriate approach for each material. He reported that using non-linear elastic approach and dividing the cold mix layer into sub-layers, each with a unique stiffness, results in higher stress ratios which are better estimates than those obtained by linear elastic calculations. This observation clearly corroborates the suggestions of Huang (2004) on same issue. Therefore the non-linear elastic approach was adopted in this exercise.

4   Non-Linear Elastic Analysis

The KENLAYER mode in the KENPAVE computer program developed by Huang (2004) for flexible pavement design which is widely available was found useful in this regard. Huang (2004) reported that the backbone of KENLAYER is the solution for an elastic multilayer system under a circular area. The solutions are superimposed for multiple wheels, applied iteratively for non-linear layers, and collocated at various times for viscoelastic layers. Therefore, KENLAYER is deemed capable of analysing layered systems under single, dual, dual-tandem or dual-tridem wheels with each layer behaving differently, linear elastic, non-linear elastic or viscoelastic.

For non-linear elastic analysis in KENLAYER, the K-θ model described in Oke (2010) applies. Huang (2004) stated that 3 methods have been incorporated into KENLAYER for non-linear analysis. While Method 1, the most accurate but time consuming (Huang, 2004) and in which the stress dependent layer is subdivided, was used in this work, in light of the suggestions of Ebels (2008) for similar materials, the other 2 methods and detailed description of the capabilities of the program are documented by Huang (2004).

The K_1, and K_2 values obtained from the resilient modulus test reported by Oke (2010) which are required for this exercise are detailed in Table 2. The other relevant required properties of the pavement materials are similarly detailed in Table 3. The 'k' i.e. K_1, K_2, K_3, K_4 values are material constants (Huang, 2004). Ekwulo and Eme (2009) noted that K_1 and K_2 are dependent on moisture content while K_3 and K_4 are related to the soil types, either coarse grained or fine grained. While all layers are assumed to be fully bonded, the following medium condition parameters (see Figure 2) suggested by Huang (2004) were also used in the analysis for the subgrade and subbase layers:

- Coefficient of Earth Pressure = K_0 = 0.8
- Subbase K_1 = 30.912
- Subgrade K_1 = 52.992 (medium condition)
- K_2 = 42.8
- K_3 = 1110
- K_4 = 178
- Emin – Emax = 30000 – 100000 (minimum and maximum stiffness modulus for subgrade)

These values were used because materials for the layers in question were not studied in this work. Meanwhile, for the surfacing layer on top of the CBEM, HMA Stiffness Modulus of 2500MPa was used. This was adopted and made constant for all cases considered since from initial trials conducted though not reported in this thesis for brevity, changes in stiffness modulus of such layers do not have significant effect on the strain and stress distribution in the underlying layers although the results indicated that as stiffness modulus of the surfacing layer increases, fatigue life increases.

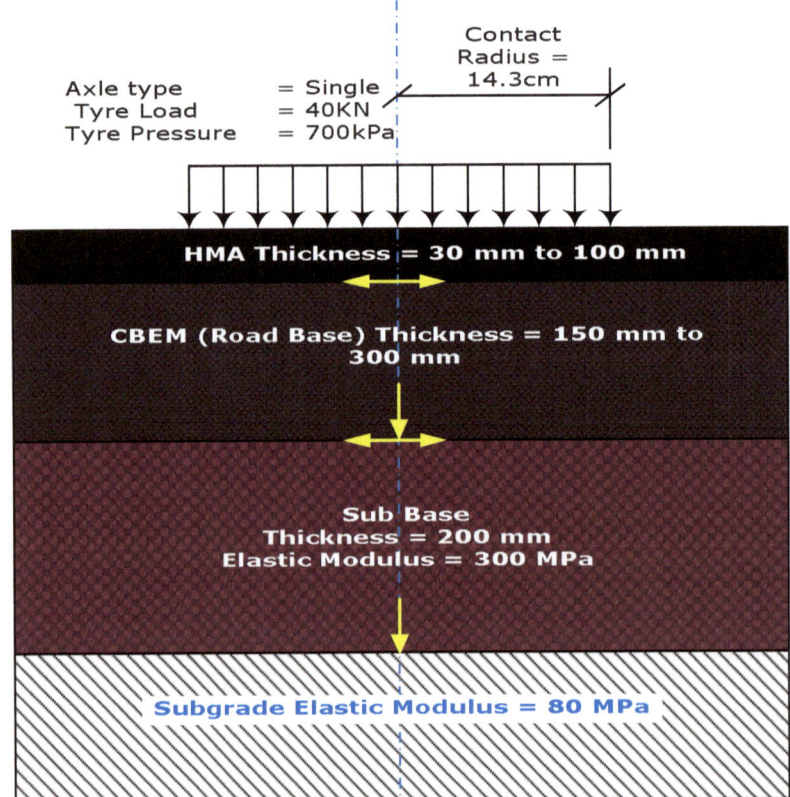

Figure 1: Hypothetical Road Pavement Structure Containing CBEMs used for the Modelling

Table 2: K_1, K_2 and R^2 Values from Resilient Modulus Test of CBEMs

CBEM Type	FC @ 20°C			FC 30°C		
	K_1 (MPa)	K_2	R^2	K_1 (MPa)	K_2	R^2
VA	26.467	0.5453	0.9885	50.026	0.436	0.9878
5dmm	50.426	0.4673	0.9771	78.226	0.378	0.9827
10dmm	32.603	0.5248	0.9895	59.325	0.4416	0.9838
20dmm	35.731	0.5319	0.9821	48.567	0.4641	0.992

FC: Fully Cured

Table 3: Pavement Layer Material Properties

Layer Material	Density (kg/m³)	Poisson's Ratio	Elastic Behaviour
HMA	2400	0.35	Linear
VA	2180	0.35	Non-Linear
5dmm	2120	0.35	Non-Linear
10dmm	2158	0.35	Non-Linear
20dmm	2224	0.35	Non-Linear
Sub base	2000	0.35	Non-Linear
Subgrade	1800	0.35	Non-Linear

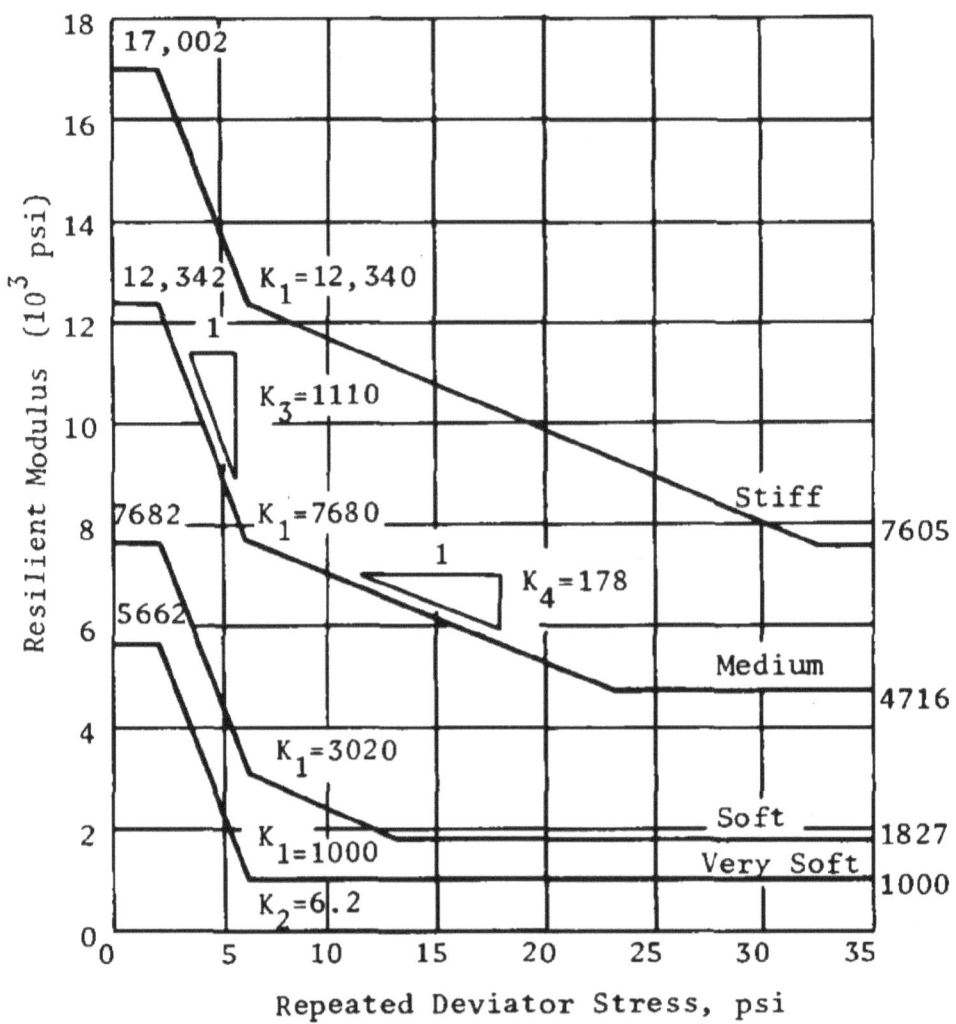

Figure 2: Resilient Modulus-Deviator stress Relationship for Four Types of Subgrade (1 psi = 6.9 kPa) (Thompson and Elliot in Huang, 1994, pg 109)

Seeding resilient modulus (compressive modulus) values for CBEMs (non-linear and stress dependent layer) ranged between 1000MPa and 600MPa depending on the number of sub-layers (each sub-layer of the CBEM is 50mm thick). This seeding value is needed because an iterative procedure is used, in which the moduli of non-linear layers are adjusted as the stresses vary, while the moduli of the linear layers remain the same. During each iteration, a constant set of moduli is computed based on the stresses obtained from the previous iteration. After the stresses due to single or multiple wheels are determined, the elastic moduli of non-linear layers are recalculated and a new set of stresses is determined. This is repeated until the moduli converge to a specified tolerance (Huang, 2004). The Sub-base Modulus was taken as 300MPa (stabilised laterite of 30% CBR) and 80MPa (lateritic soil of 8% CBR) was adopted for the Subgrade Modulus since these were observed for such layers for a typical Nigerian road by Ekwulo and Eme (2009). The other required information such as loads and layer thicknesses are as stated in Figure 1 and Table 1.

It is worth noting that all required entries must be made correctly as the KENLAYER program is highly sensitive to inputs. The outputs from the analysis are the vertical, horizontal and shear strains and stresses. Also included is resilient modulus of each layer and vertical displacements.

It is also worth noting that although this multi-layer analysis is easy to execute (Oliveira, 2006), researchers have found out that performance prediction from such an approach is not as accurate as those obtained from finite element analysis (Oliveira, 2006; Huang, 2004; Ebels, 2008; and Twagira, 2010). However, Oliveira (2006) and Twagira (2010) both observed that finite element analysis is difficult to implement unless users have a thorough knowledge of basic principles.

5 Stress and Strain Distribution in Pavements Containing CBEMs

Using pavement structure case number 9 from Table 1 i.e. surfacing layer thickness of 30mm, CBEM (road base) layer thickness of 250mm (divided into 5 equal sub-layers) and a sub-base layer of 200mm, the stress and strain distributions through pavements containing VACBEM, 5dmmCBEM, 10dmmCBEM and 20dmmCBEM as road base layers which were analysed using the KENLAYER analytical tool were studied. All the CBEMs used were in a fully cured condition but tested at 20°C and 30°C.

Figures 3 to 6 are relevant here. Figure 3 shows the vertical stress distribution in the pavements. The results of the analyses indicate that vertical stress distribution in the CBEMs is not significantly different for all the pavement types considered. The figure shows also that vertical stress reduces with depth. These were also observed for the horizontal stress distribution as indicated in Figure 4 though with more pronounced differences indicated in the area covered by the red dash line circle. It is to be noted that the differences here occurred in the CBEM layer only since all other layers have similar properties for all the pavements considered. The figure also shows that the horizontal stress gradually changed with depth from a relatively high negative value at the top of the surfacing layer to a positive value on top of the subgrade. Figure 5 shows the vertical strain distribution in the pavements while Figure 6 shows similar distribution for the horizontal strain in the pavements. The red dash lines in the figures show the bottom level of the CBEM layers in the hypothetical pavement structure. The figures indicate significant differences in strains for the CBEM layers and that the maximum horizontal and vertical strains are not at the bottom of the CBEM layers, but very close to the middle of the CBEM layers. These observations are consistent with those observed by Ebels (2008) for similar materials. He observed that cold mix materials with different properties showed significant differences in strains while stress values had little differences using the KENLAYER analytical tool. The figures showed that the RAP CBEMs generally have lower strain values compared to the VACBEM.

Meanwhile, as observed from the relevant figures for strains, VACBEM indicated the highest values (worst performance) for the two temperatures while the 20dmmCBEM tested at 20°C indicated the lowest strain values. Materials tested at 30°C generally showed higher strain values compared to those tested at 20°C.

6   The Effect of Analytical Method on Stress and Strain Distributions

The effect of the applied analytical tool on the stress and strain distribution in the pavement was briefly examined. This was done by comparing the stress and strain distributions generated from the KENLAYER analytical tool with those generated using the Bitumen Stress Analysis in Roads (BISAR 3.0) software which was developed by Shell (1998).

Since in the preceding section, the analyses showed that the vertical stress distributions for all the pavements considered were close, the corresponding modulus values (obtained iteratively) to the stress values observed at the bottom

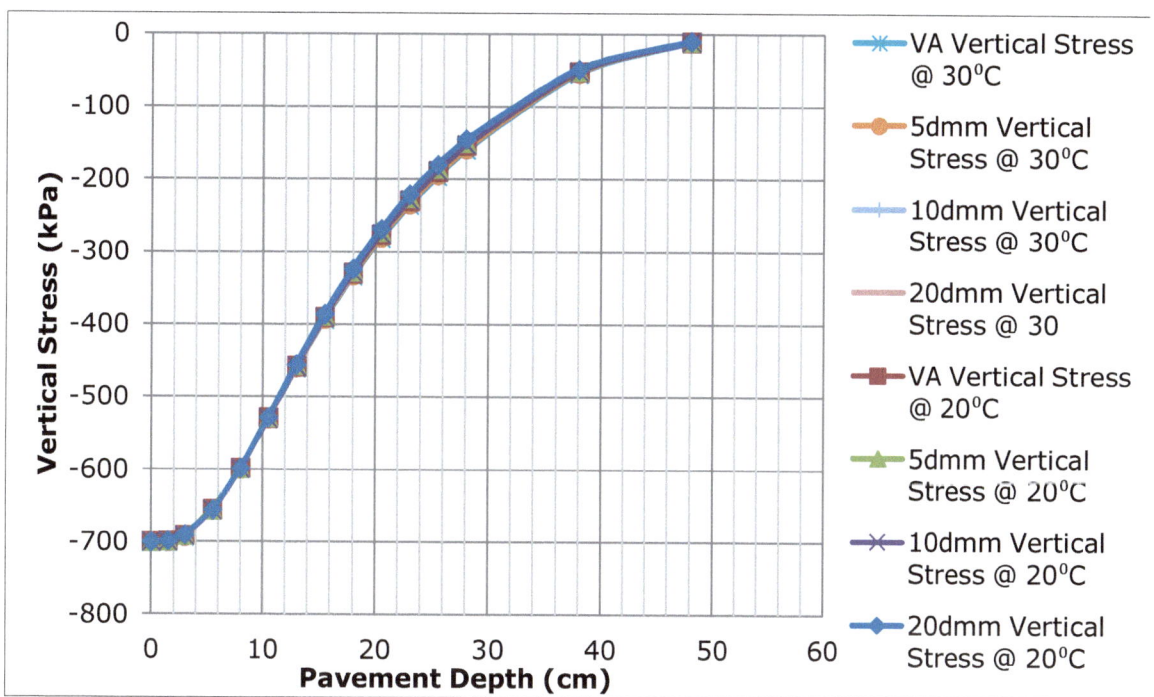

Figure 3: Vertical Stress Distribution through Pavement Structure Case Number 9 using KENLAYER

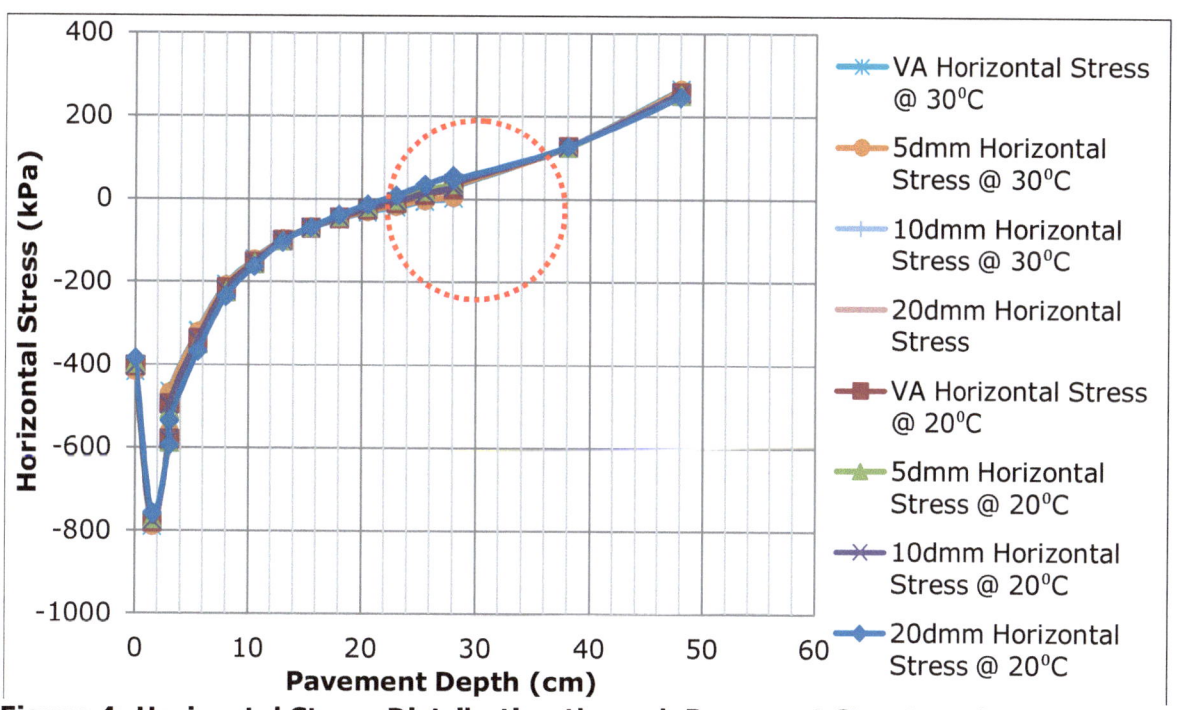

Figure 4: Horizontal Stress Distribution through Pavement Structure Case Number 9 using KENLAYER

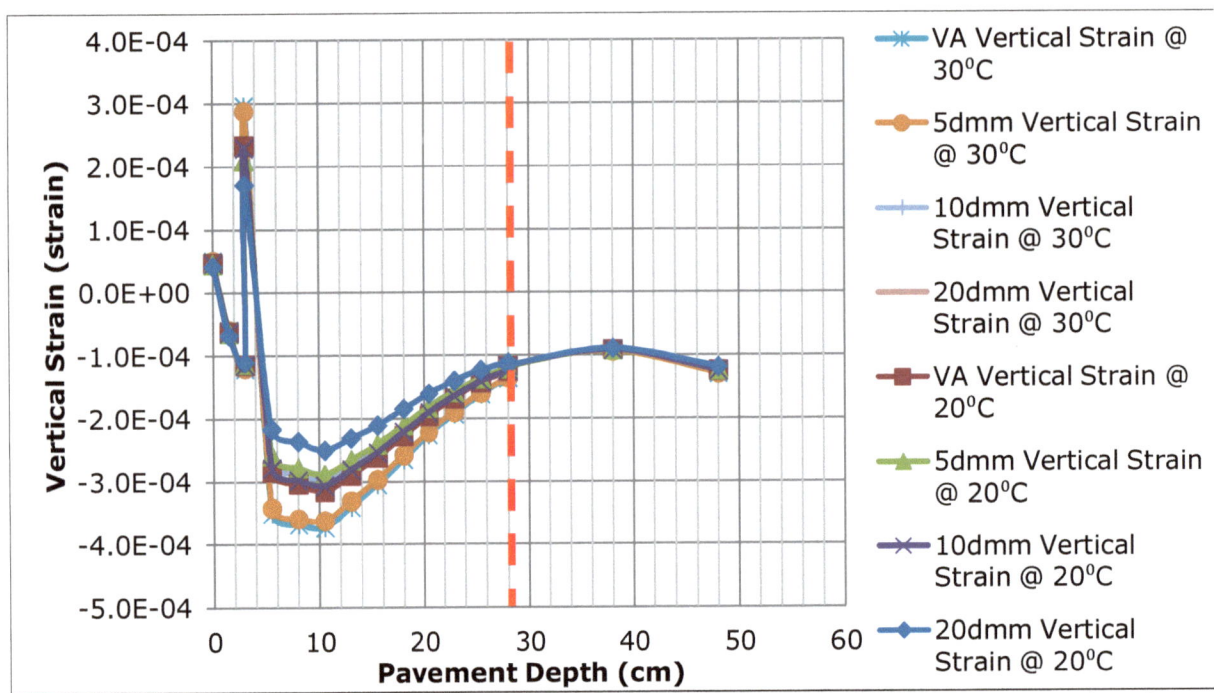

Figure 5: Vertical Strain Distribution through Pavement Structure Case Number 9 using KENLAYER

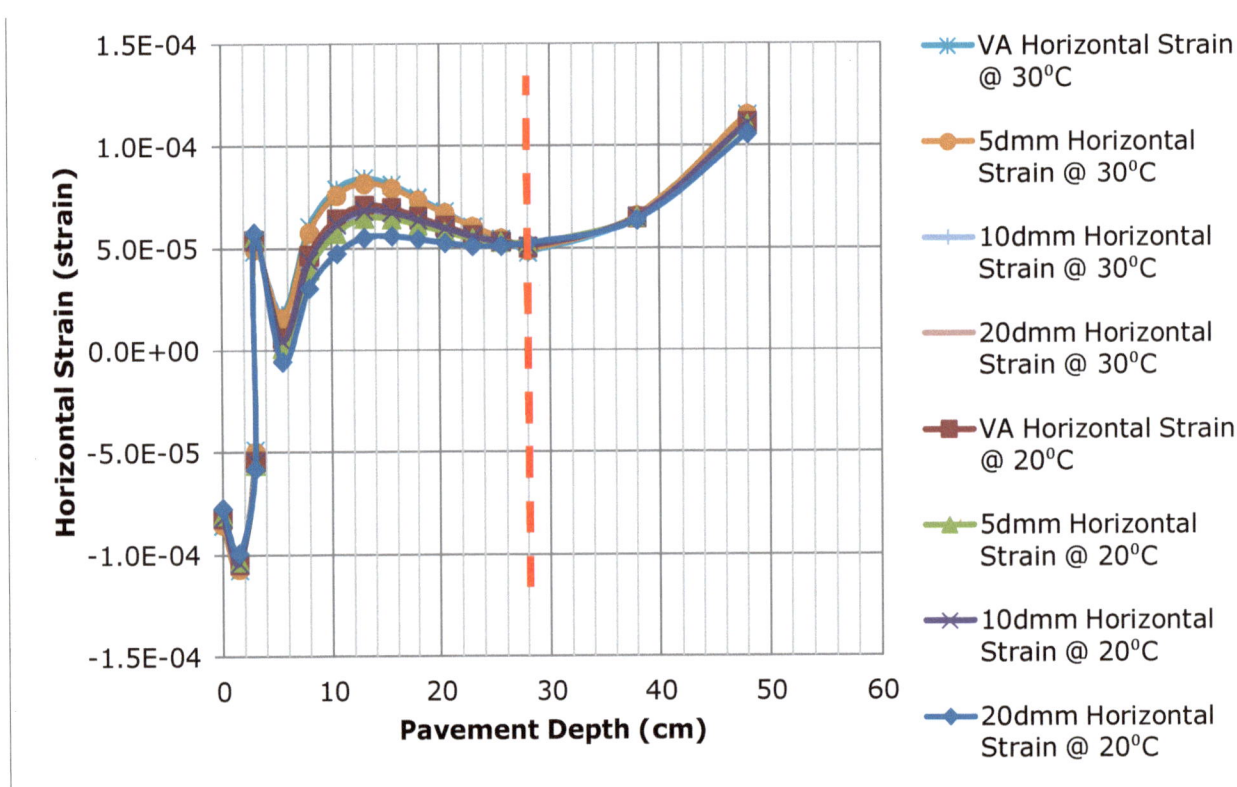

Figure 6: Horizontal Strain Distribution through Pavement Structure Case Number 9 using KENLAYER

of each of the CBEM (road base) layers were used in BISAR 3.0 (linear elastic analysis). For this case, the pavement structure number 6 i.e. surfacing layer thickness of 50mm, road base layer (CBEM) of 200mm (sub divided into 4 equal layers) and for the sub-base layer, a thickness of 200mm was used. For brevity, only the materials tested at 20°C were considered.

Figures 7 to 10 are relevant here. Figure 7 shows the vertical stress distribution for pavements containing the CBEMs using both linear and non-linear analytical tools i.e. BISAR 3.0 and KENLAYER. The figure shows that there is a little difference in the trend of stress distributions recorded by BISAR 3.0 and KENLAYER although the two results indicated no significant differences between pavement types considered within each of the two groups i.e. BISAR 3.0 analysis results and KENLAYER analysis results respectively. Similarly, the results of the KENLAYER analysis in the referred figure when compared with those in Figure 3 indicated that change in thicknesses of the surfacing and road base (CBEM) layers did not introduce any significant difference in stress distributions in the grouped pavements.

A not so similar trend was observed in Figure 8 which details the results of the horizontal stress distributions for the BISAR 3.0 and KENLAYER analyses with a significant difference in the stress distribution in the surfacing layer (HMA). The BISAR 3.0 results indicated very high compressive stresses (about 2000kPa) on top of the HMA layer and similarly high tensile stresses (about 1000kPa) at the bottom of the same layer. The KENLAYER recorded for this same pavement little compressive stresses i.e. about 400kPa and 500kPa on top and at the bottom of the HMA respectively. These obviously are responsible for the trends observed for the strains in Figures 9 and 10 with BISAR 3.0 returning large strain values.

Although as expected, the stress distributions converged at the bottom of the CBEM layers since it was the vertical stresses at such points that were used to calculate the corresponding modulus values used in the BISAR 3.0 analysis, the observed trends clearly show that BISAR 3.0 must have overestimated the vertical and horizontal strains required for design. Such high strain values will affect the sizing of the pavement layers which might be practically uneconomical. The modulus values used in the BISAR 3.0 analyses for the CBEM layers were generally low (about 300MPa) and this strongly affected the response of the surfacing (returning very large strain values).

Since modulus values were calculated for each of the sub-layers of the CBEMs in KENLAYER, another set of analyses was conducted in BISAR 3.0 for the same

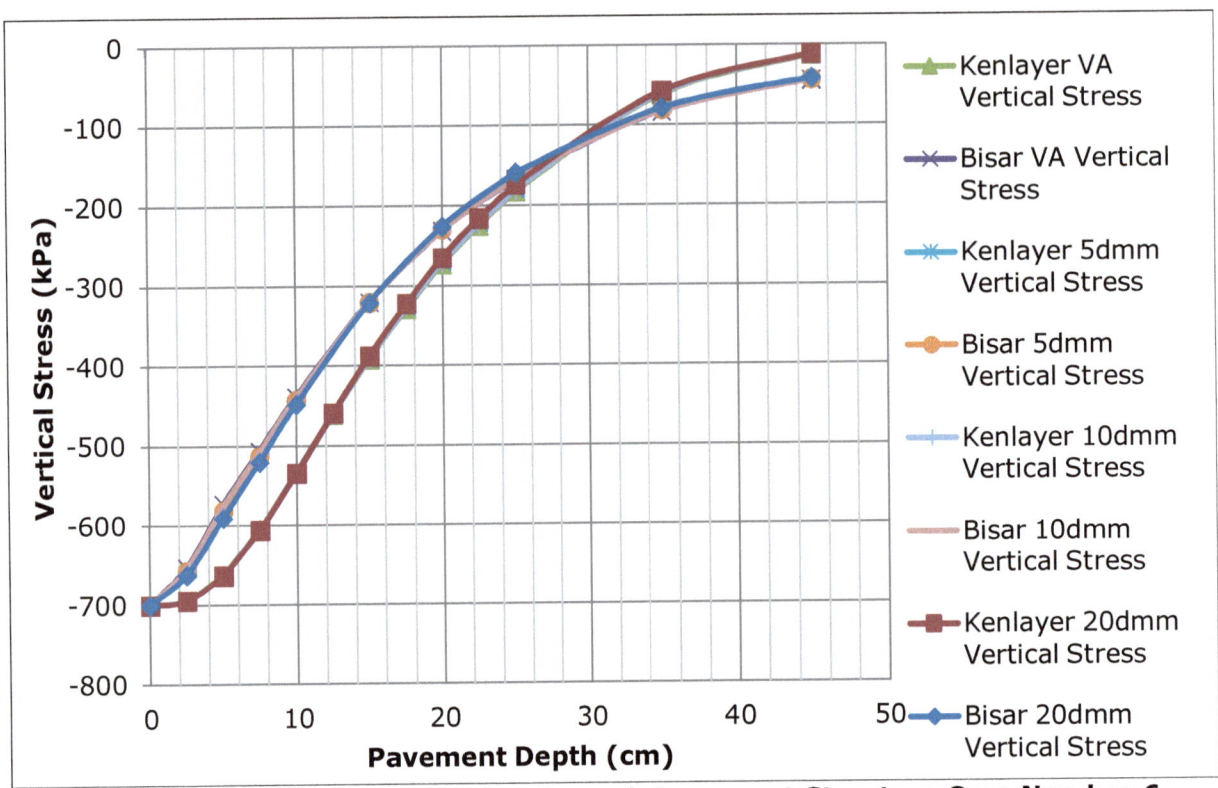

Figure 7: Vertical Stress Distribution through Pavement Structure Case Number 6 using BISAR 3.0 and KENLAYER

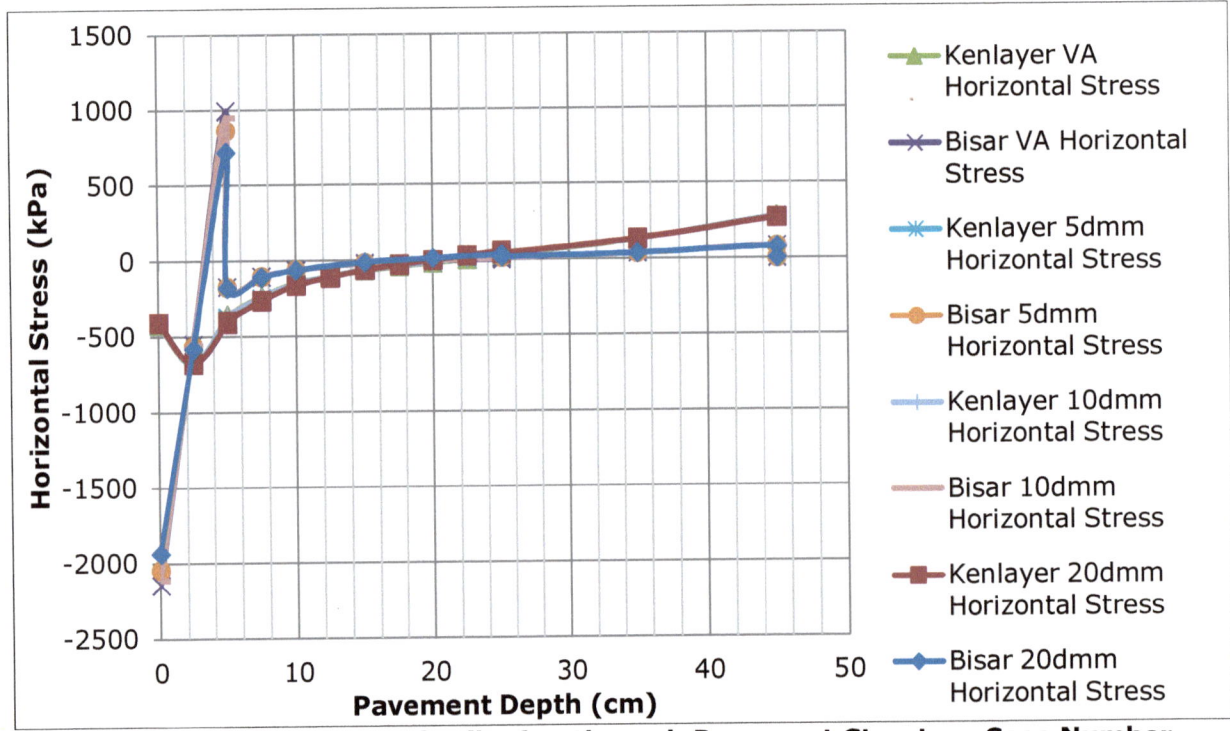

Figure 8: Horizontal Stress Distribution through Pavement Structure Case Number 6 using BISAR 3.0 and KENLAYER

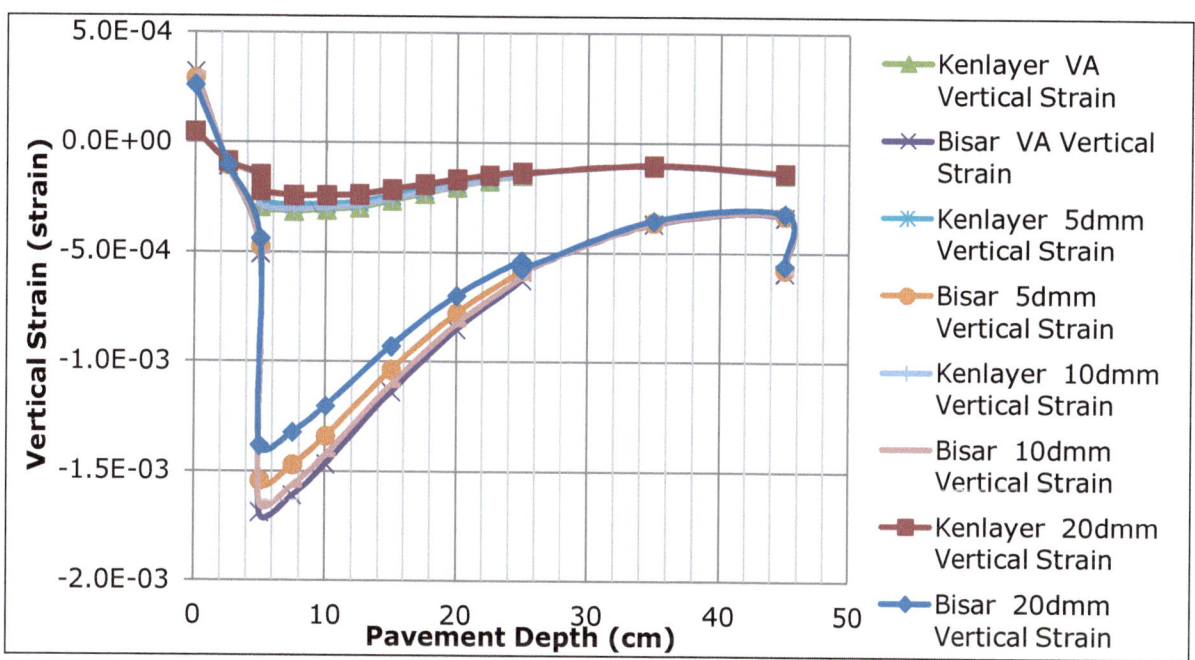

Figure 9: Vertical Strain Distribution through Pavement Structure Case Number 6 using BISAR 3.0 and KENLAYER

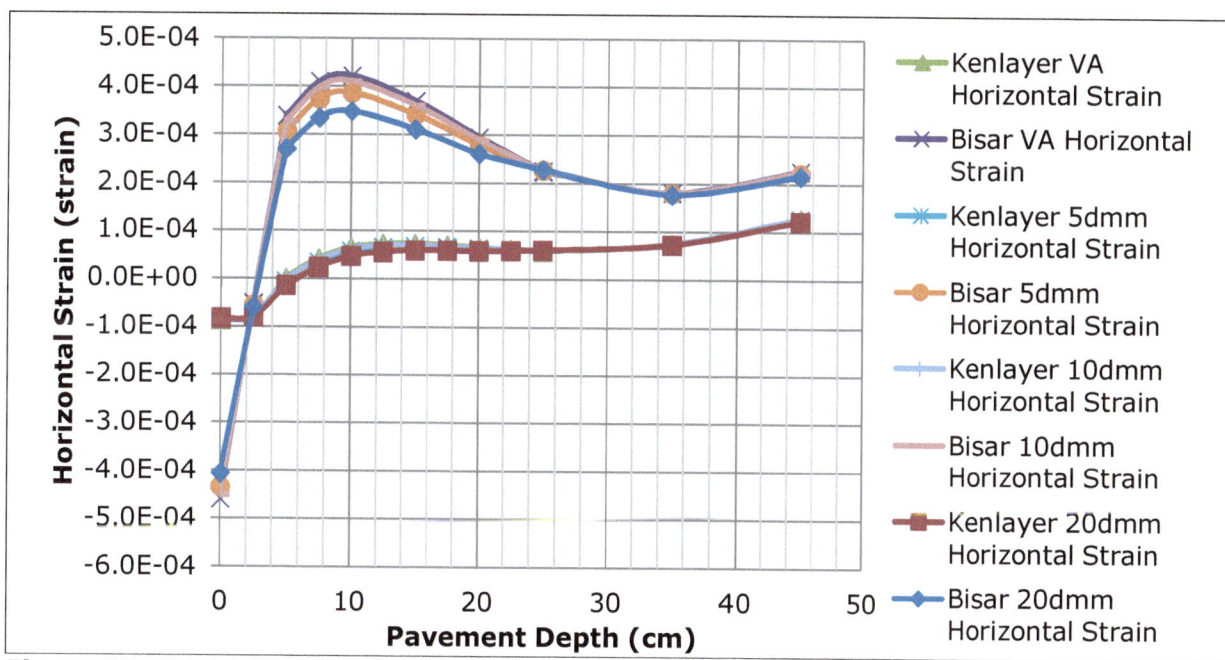

Figure 10: Horizontal Strain Distribution through Pavement Structure Case Number 6 using BISAR 3.0 and KENLAYER

pavement now using the averages of the calculated modulus values for each of the CBEM layers from KENLAYER. Table 4 details the modulus values. It was observed from the table that modulus reduces from top to bottom which reflects the stress dependency of the

CBEMs. Also since the materials were stress dependent, the stress values at the bottom of the CBEMs which were used for the first set of analyses on the BISAR 3.0 were not large enough to represent the average characteristic properties of the CBEMs.

Figures 11 to 14 indicate that when the correct modulus values of the sub-layers of the CBEMs are known, BISAR 3.0 can also be used to determine the stress and strain distributions through the non-linear layers just as the KENLAYER does. This observation is consistent with the findings of Ebels (2008). The sub-layers are now treated as linear layers in this regard. However it is of no use using BISAR 3.0 to estimate the stress and strain distributions in such pavements containing non-linear layers since estimating the required modulus values needed for accurate estimates of strains and stresses in BISAR 3.0 is first done using KENLAYER. It is better to execute such designs involving non-linear layers with KENLAYER. BISAR 3.0 therefore might not be suitable for analysing pavements containing such non-linear materials.

7   Design of CBEM Road Base for Fatigue Life

The results of the analyses done in the preceding sections using KENLAYER showed that the maximum horizontal strains in the CBEM layers are somewhere close to the middle, and thus, such strains are deemed more appropriate theoretically for design purposes than those observed at the bottom of the CBEM (road base) layers recommended by Ebels (2008). The design presented here considered these set of strains i.e. maximum and at the bottom, to ascertain their suitability or not.

Where applicable, only laboratory test results (fatigue line equations reported in Oke (2010)) conducted at 30°C were used since the focus of the work is hot tropical climate. A conventional HMA road base layer of 28mm DBM (Read, 1996) though analysed using BISAR 3.0 was included for comparisons. A stiffness value of 3000MPa was assumed for the layer (at 30°C). The design was conducted for both critical condition (point at which pavement crack initiation occurs) and failure. The shift factors of 77 and 440 were used for the respective points as suggested by Brunton (1983). It is worth noting however that such shift factors might not be appropriate for the CBEMs (cold mixtures) since they were developed for HMA. At present, no such universally accepted values are available for cold mixtures.

Table 4: Calculated Modulus Values for each CBEM Sub-Layer in KENLAYER (@ 20°C)

Sub-layer No.	Modulus Values from KENLAYER Analyses			
	VA CBEM	5dmmCBEM	10dmmCBEM	20dmmCBEM
1	1400	1519	1431	1746
2	1308	1421	1350	1614
3	1296	1399	1335	1561
4	1324	1415	1457	1596
Average Modulus	1332	1439	1368	1634

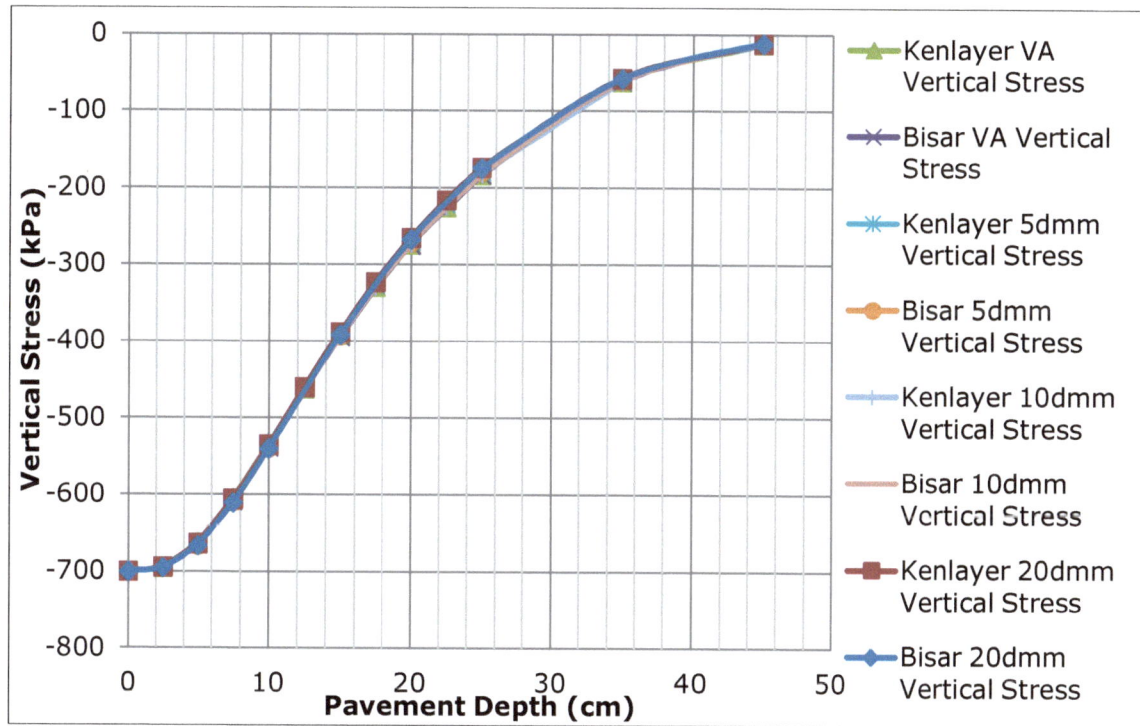

Figure 11: Vertical Stress Distribution through Pavement Structure Case Number 6 using BISAR 3.0 (with calculated modulus) & KENLAYER (@ 20°C)

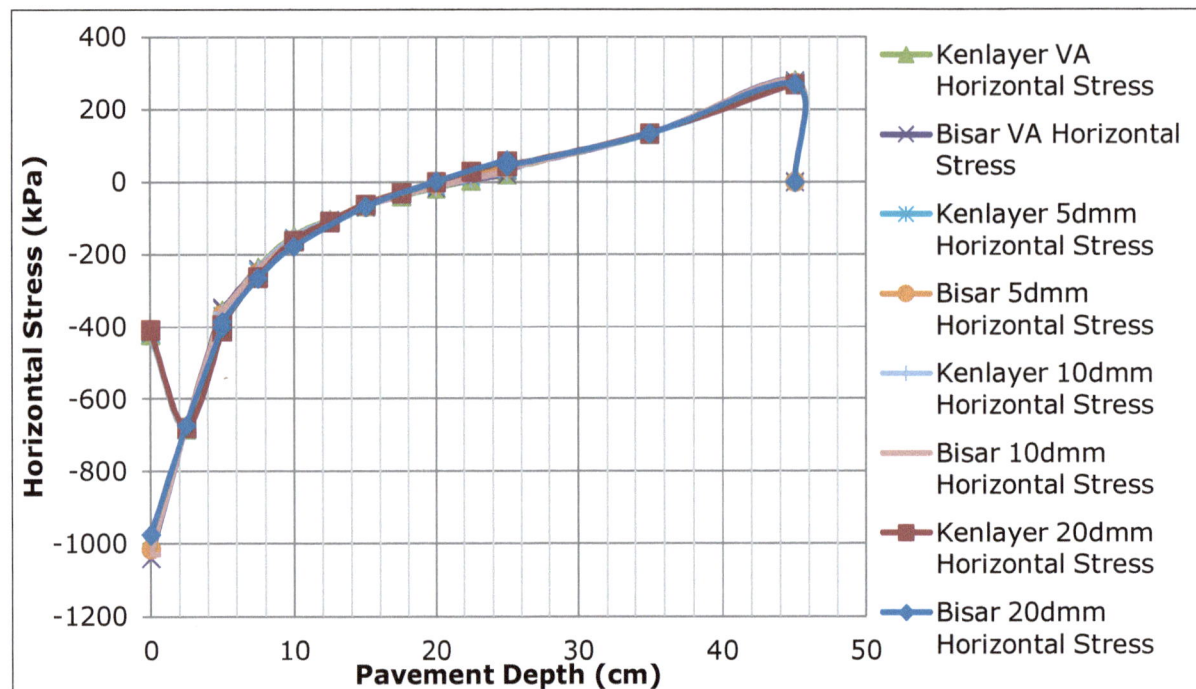

Figure 12: Horizontal Stress Distribution through Pavement Structure Case Number 6 using BISAR 3.0 (with calculated modulus) & KENLAYER (@ 20°C)

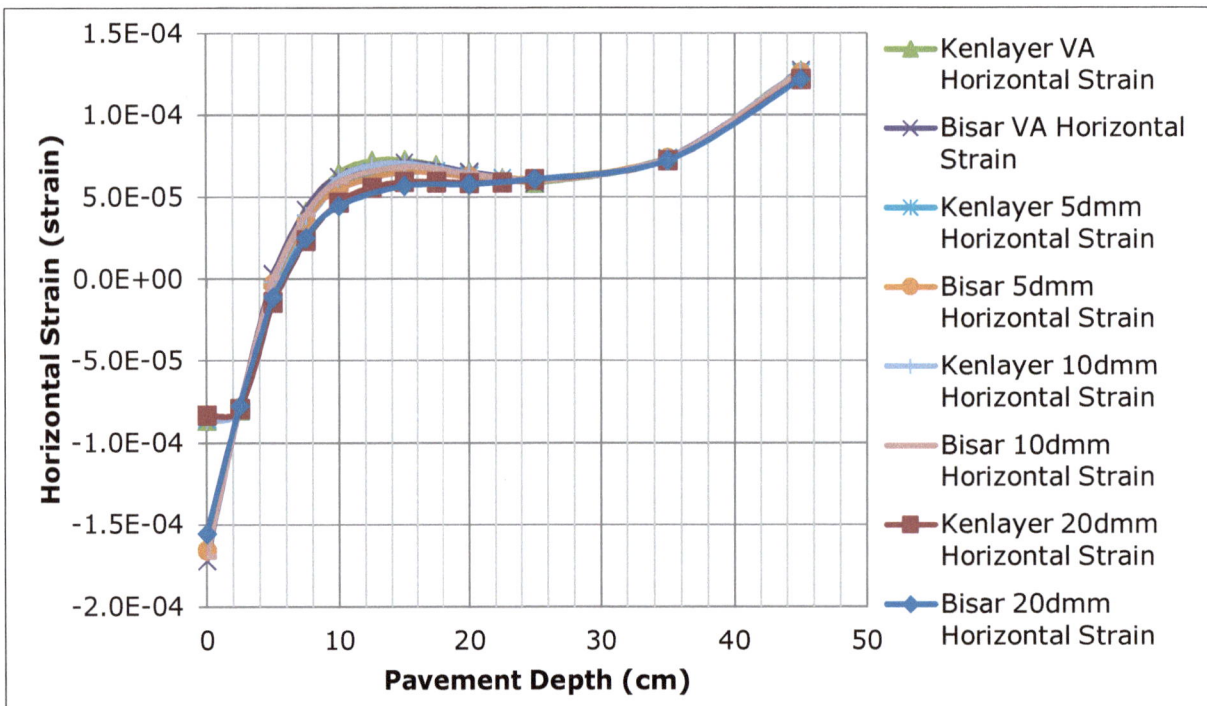

Figure 13: Horizontal Strain Distribution through Pavement Structure Case Number 6 using BISAR 3.0 (with calculated modulus) & KENLAYER (@ 20°C)

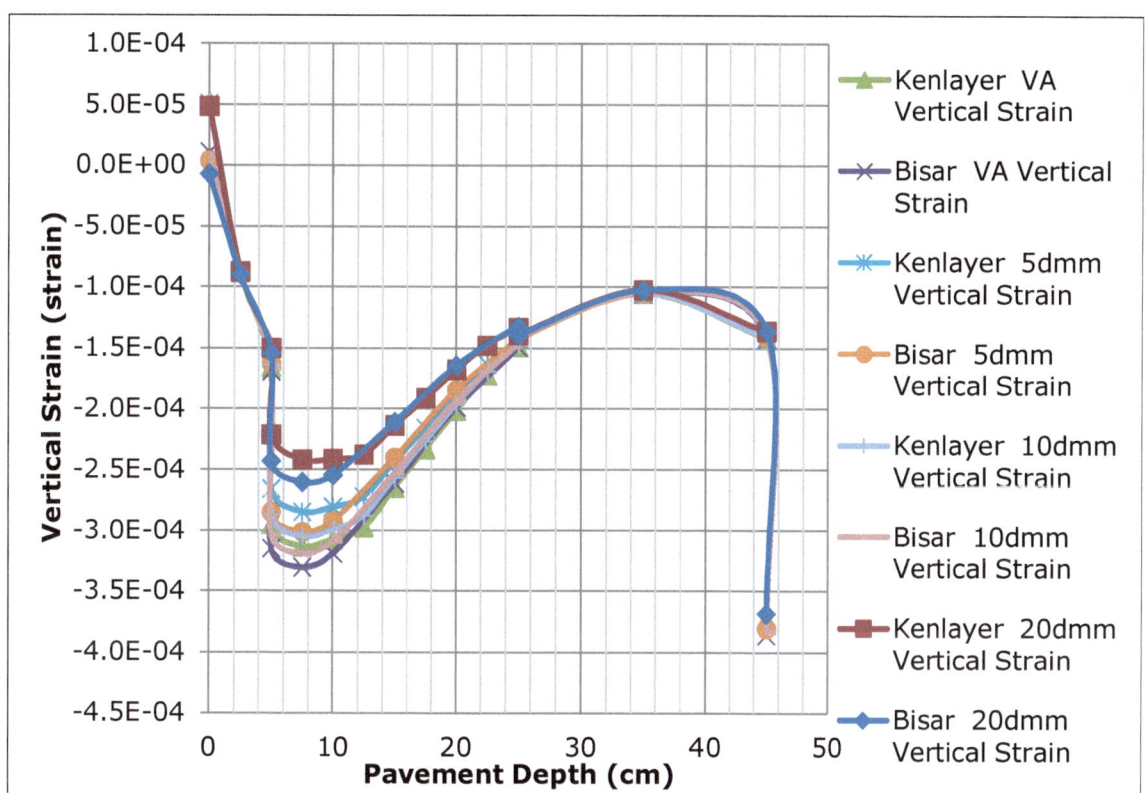

Figure 14: Vertical Strain Distribution through Pavement Structure Case Number 6 using BISAR 3.0 (with calculated modulus) and KENLAYER (@ 20°C)

While the 440 value for life to failure might be too high, 77 may be too low to describe total pavement life. Considering the fact that cold mixtures are not as good as HMA as observed in this thesis, it is believed that the appropriate value for life to failure should be somewhere in-between these two values. Thus, the results presented here are just estimates and can only be used with caution. Figures 15a to 17 detail design charts based on horizontal strains at the bottom of the CBEM layers and are the results for equivalent standard axle loads (ESALs) to critical condition. Figures 18 to 21 show design charts based on maximum horizontal strains in the CBEM layers and are similarly the results for ESAL to critical condition. The surfacing layer thicknesses used were 30mm, 50mm, 75mm and 100mm. The road base layer thicknesses used were 150mm, 200mm, 250mm and 300mm.

For cases where horizontal strains at the bottom of the CBEM layers were used, the results clearly indicate that as CBEM layer thickness increases, fatigue lives (to critical or failure point) increase. This was similarly observed for the HMA 28mm DBM layer although with better responses in life to thickness increases. The results also indicate that changes in the surfacing layer thickness do not cause significant improvement to the fatigue lives of the

CBEM road base layers compared to the HMA. Of all the CBEM materials considered for the road base layer, the 5dmmCBEM had the best performance while the VACBEM had the worst performance. All the RAP CBEMs and the HMA 28mm DBM had fatigue lives clearly above 1 million ESALS applications for failure and 0.1 million for critical condition irrespective of the layer thickness.

Except for ranking the CBEMs in the same order as observed for horizontal strains at the bottom of the CBEM layers, the results for using maximum horizontal strains in the CBEM layers as indicated in Figures 18 to 21 show that changes in surfacing and CBEM layers' thicknesses were of no effect on the fatigue lives of the CBEM layers. This indeed must be responsible for the earlier mentioned recommendation of Ebels (2008) that strains at the bottom be used for such design purposes since increase in thicknesses of the CBEM layers as earlier observed in Figures 15a to 17 caused an increase in their fatigue lives. Although not reported in this thesis, similar observations were noted for the rutting lives of the CBEMs using maximum vertical strains in the CBEM layers. However, rather than using the horizontal strains at the bottom of the CBEM layer for fatigue design lives which are lower than the maximum which is not common in practice, it is not advisable to use the fatigue and rutting lives of the

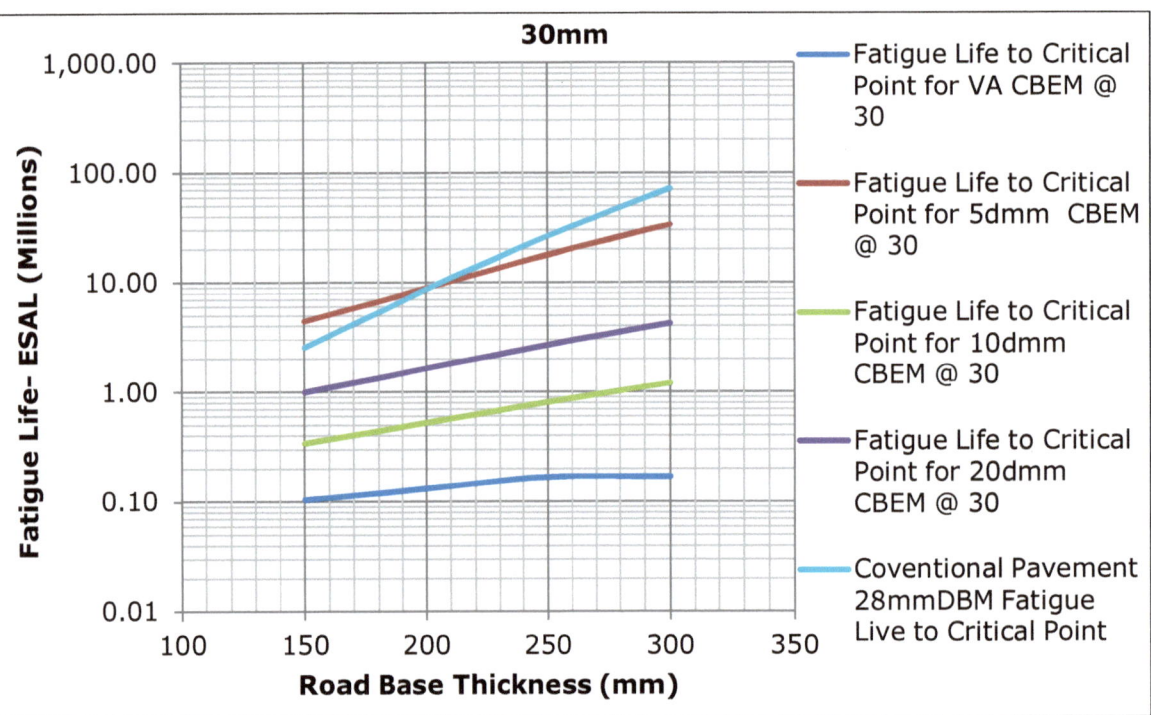

Figure 15a: Designed Road Base Layer Thicknesses Using Horizontal Strains at the Bottom of CBEM Layers to Determine Corresponding Fatigue Lives to Critical Condition for Pavements with Surfacing Thickness of 30mm HMA

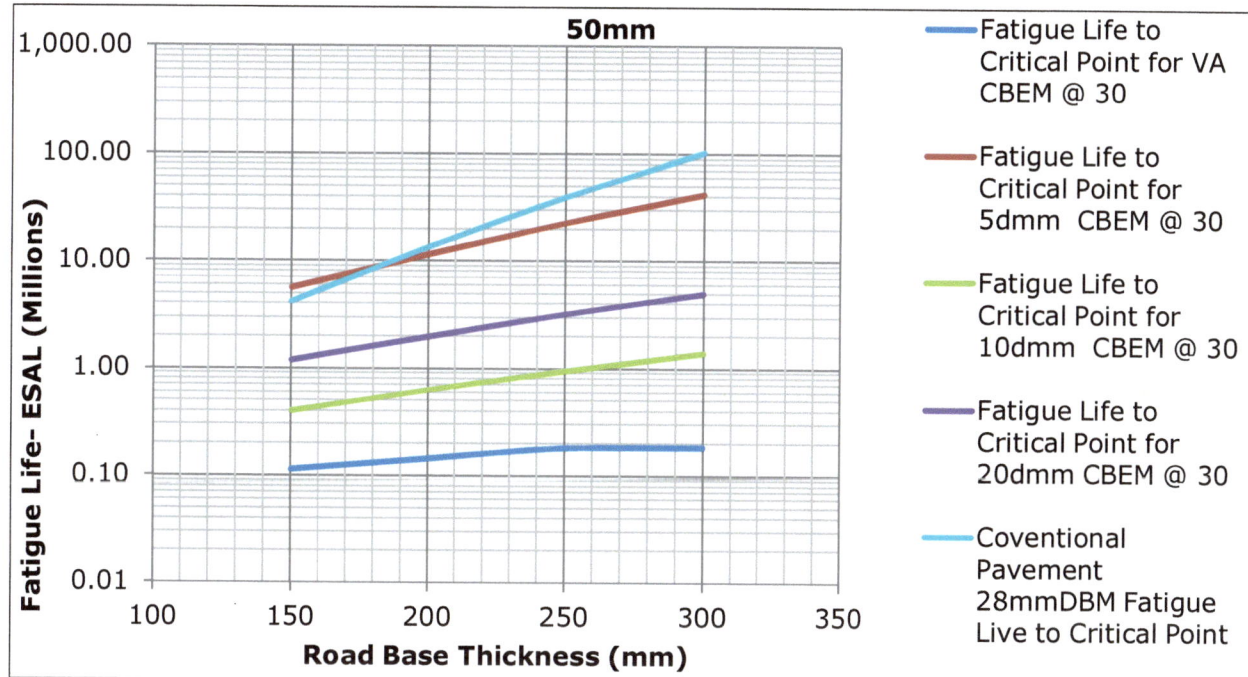

Figure 15b: Designed Road Base Layer Thicknesses Using Horizontal Strains at the Bottom of CBEM Layers to Determine Corresponding Fatigue Lives to Critical Condition for Pavements with Surfacing Thickness of 50mm HMA

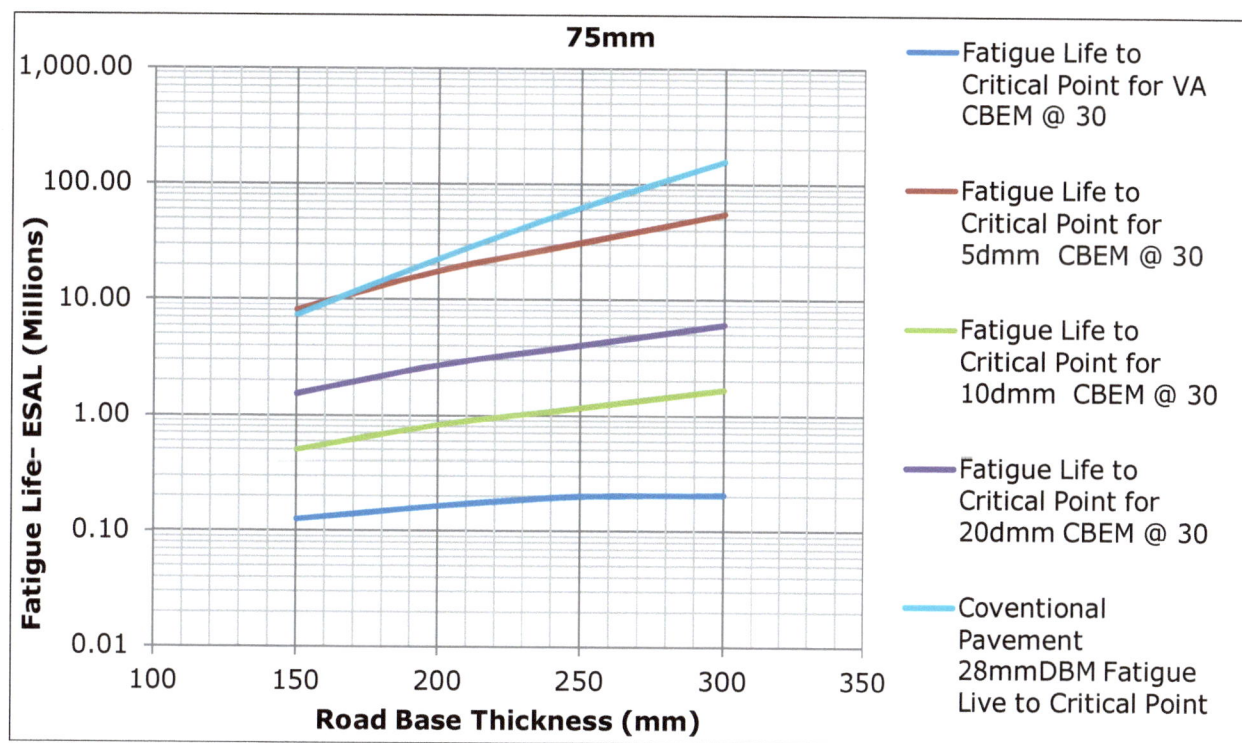

Figure 16: Designed Road Base Layer Thicknesses Using Horizontal Strains at the Bottom of CBEM Layers to Determine Corresponding Fatigue Lives to Critical Condition for Pavements with Surfacing Thickness of 75mm HMA

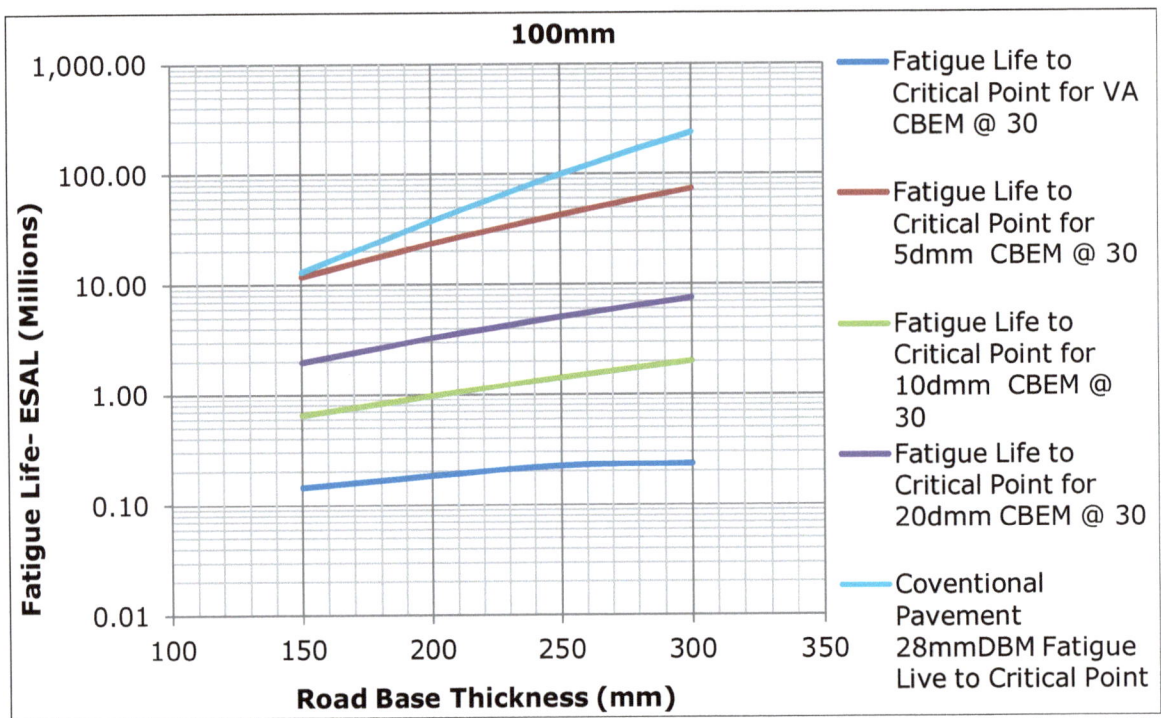

Figure 17: Designed Road Base Layer Thicknesses Using Horizontal Strains at the Bottom of CBEM Layers to Determine Corresponding Fatigue Lives to Critical Condition for Pavements with Surfacing Thickness of 100mm HMA

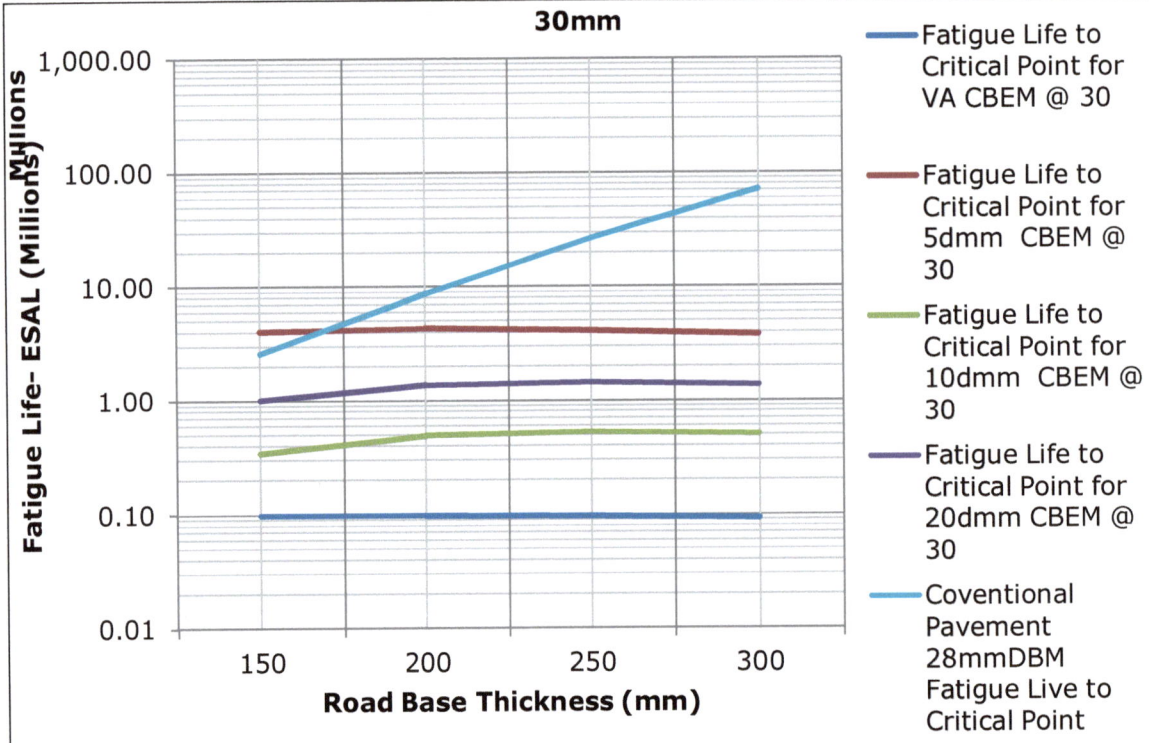

Figure 18: Designed Road Base Layer Thicknesses Using Maximum Horizontal Strains in the CBEM Layers to Determine Corresponding Fatigue Lives to Critical Condition for Pavements with Surfacing Thickness of 30mm HMA

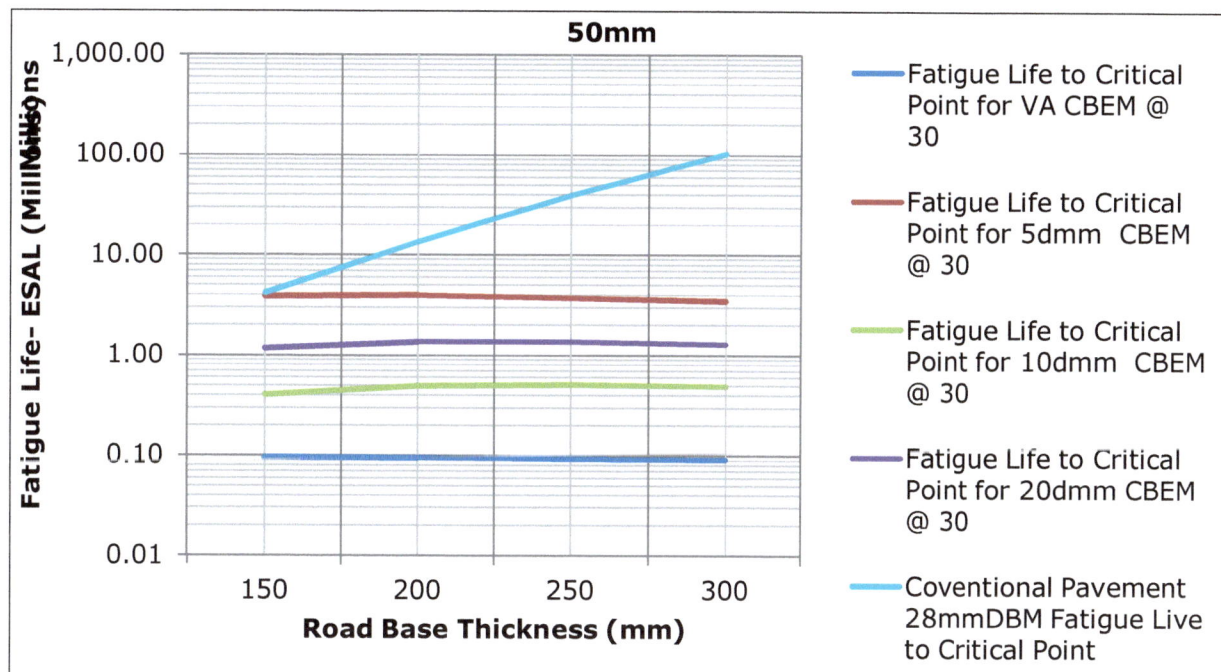

Figure 19: Designed Road Base Layer Thicknesses Using Maximum Horizontal Strains in the CBEM Layers to Determine Corresponding Fatigue Lives to Critical Condition for Pavements with Surfacing Thickness of 50mm HMA

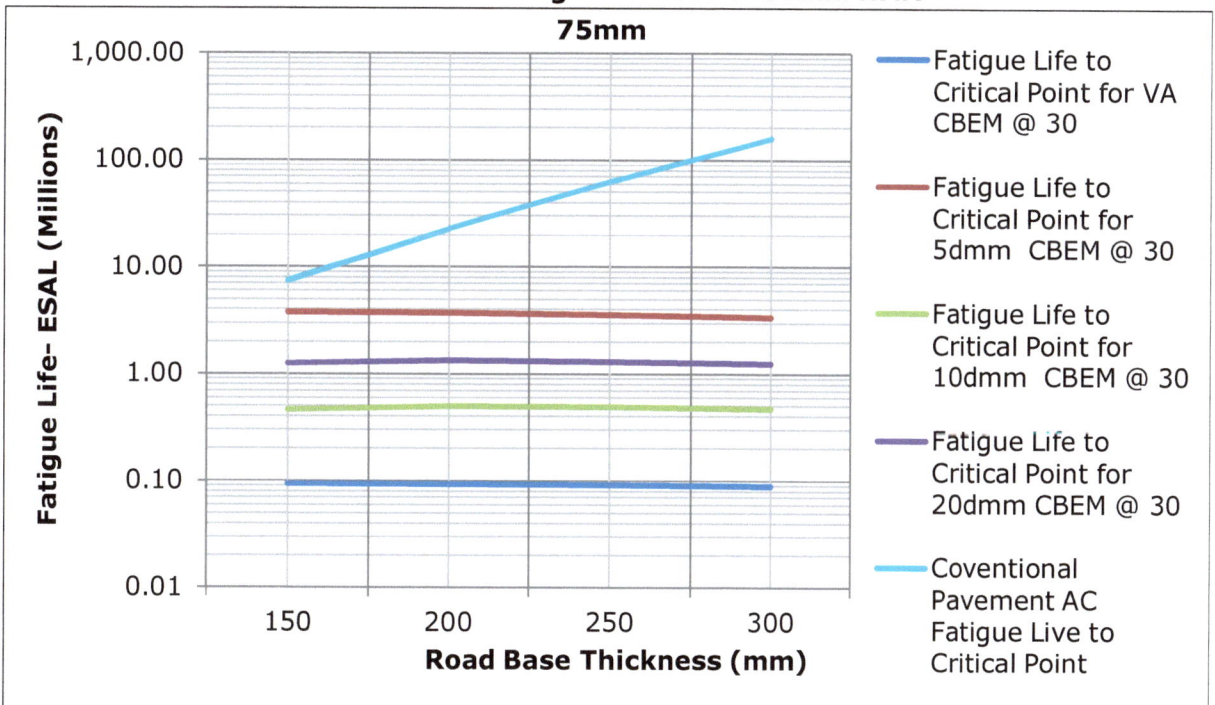

Figure 20: Designed Road Base Layer Thicknesses Using Maximum Horizontal Strains in the CBEM Layers to Determine Corresponding Fatigue Lives to Critical Condition for Pavements with Surfacing Thickness of 75mm HMA

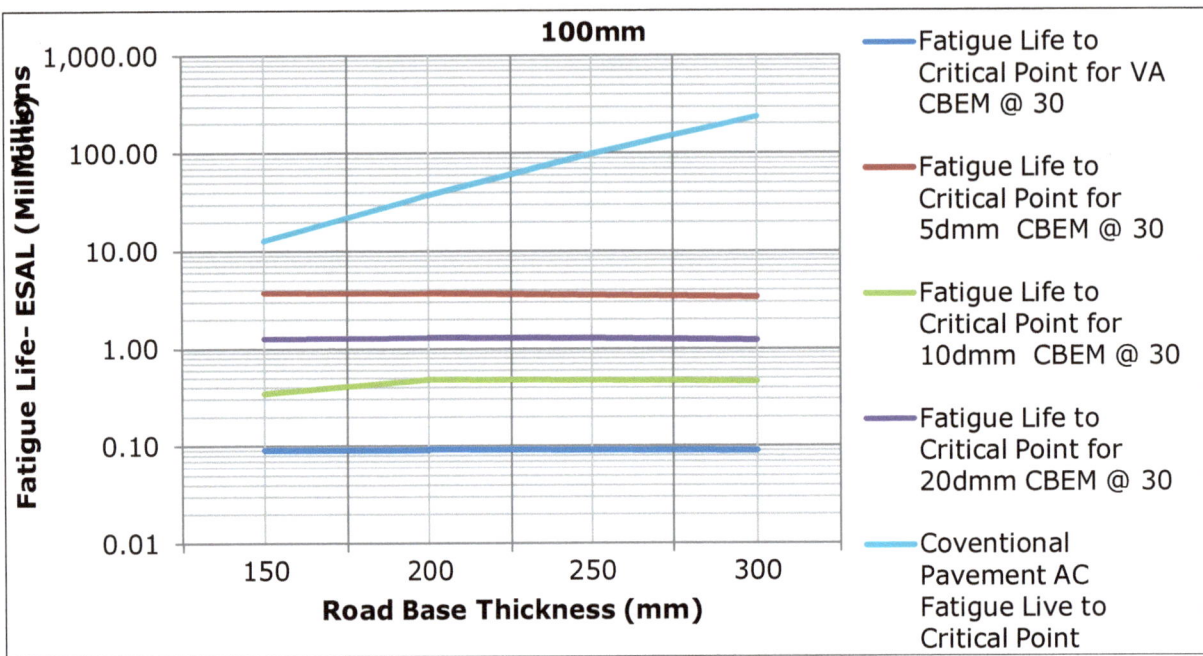

Figure 21: Designed Road Base Layer Thicknesses Using Maximum Horizontal Strains in the CBEM Layers to Determine Corresponding Fatigue Lives to Critical Condition for Pavements with Surfacing Thickness of 100mm HMA

CBEM layers as criteria for design. It is however possible that the representative strains required for considering CBEM layers in structural design of pavements can be ascertained through full scale pavement tests and this will be an interesting area for further research work.

8   The Effect of CBEM Road Base Layer Property on the Fatigue Life of Surfacing Layer

The fatigue lives of the overlying surfacing HMA layers were studied to ascertain the effect of the property of the road base (CBEM) on such. The HMA used here was the 30/14 HRA studied by Read (1996). In the initial analyses conducted using the horizontal strains at the bottom of the HMA layers from the KENLAYER results, all the strain values for HMA layers of 30mm and some of 50mm were negative (compressive) values while the others were positive (tensile). Although the observed compressive strain value(s) is an indication of good performance for the HMA surfacing layer considering the criterion used i.e. horizontal strain at the bottom of the HMA surfacing layer, in real life, it is impossible for a pavement to last forever, which the results imply. The results also suggest that failure in the pavement due to the compressive strains may be predominantly governed by top down cracking.

Therefore another analysis was conducted using OLCRACK developed by Thom (2000; and 2008). This analytical tool takes into account both the top-down and bottom-up crack propagations in the surfacing HMA layer for the estimation of standard axle loads to failure. The method of analysis in OLCRACK assumes that the layer is linear in elastic response. Since the modulus of the road base is required for such, the calculated average values for such layers done using KENLAYER were again found useful here. These were 1167MPa, 1193MPa, 1428MPa and 1390MPa for the VACBEM, 5dmmCBEM, 10dmmCBEM and 20dmmCBEM respectively at 30°C. The stiffness modulus of the HMA road base used was 3000MPa.

Figures 22 to 24 detail the results of the analysis conducted using OLCRACK. For brevity, the results for the 30mm HMA surfacing layer were not included since they showed a negative trend (between fatigue life and pavement thickness) for pavements containing CBEMs. The result suggest that the 30mm HMA surfacing is either too thin for overlay in pavements containing CBEM layers as road base or that there is a need to improve on the properties of the underlying layers i.e. sub-base and subgrade in order to obtain realistic results. Asphalt Academy (2009)

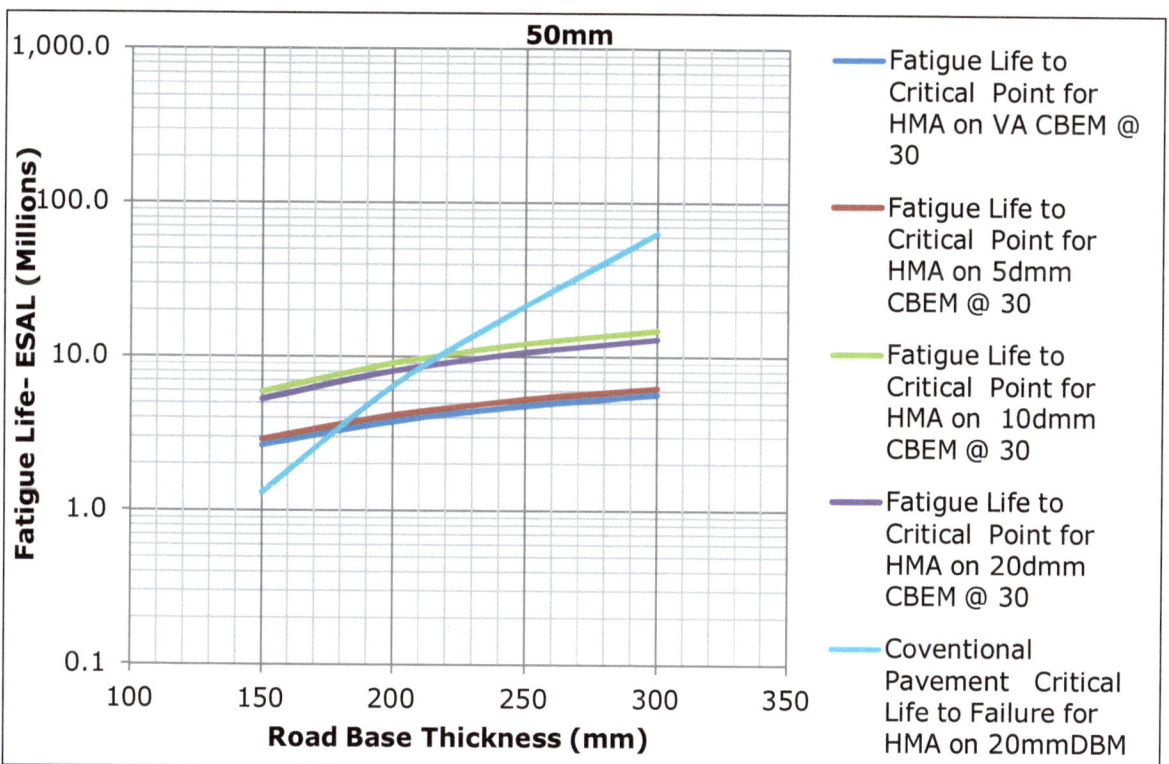

Figure 22: Effect of Road Base Layer Thickness and Material Type on Fatigue Lives to Critical Condition for Surfacing Thickness of 50mm HMA

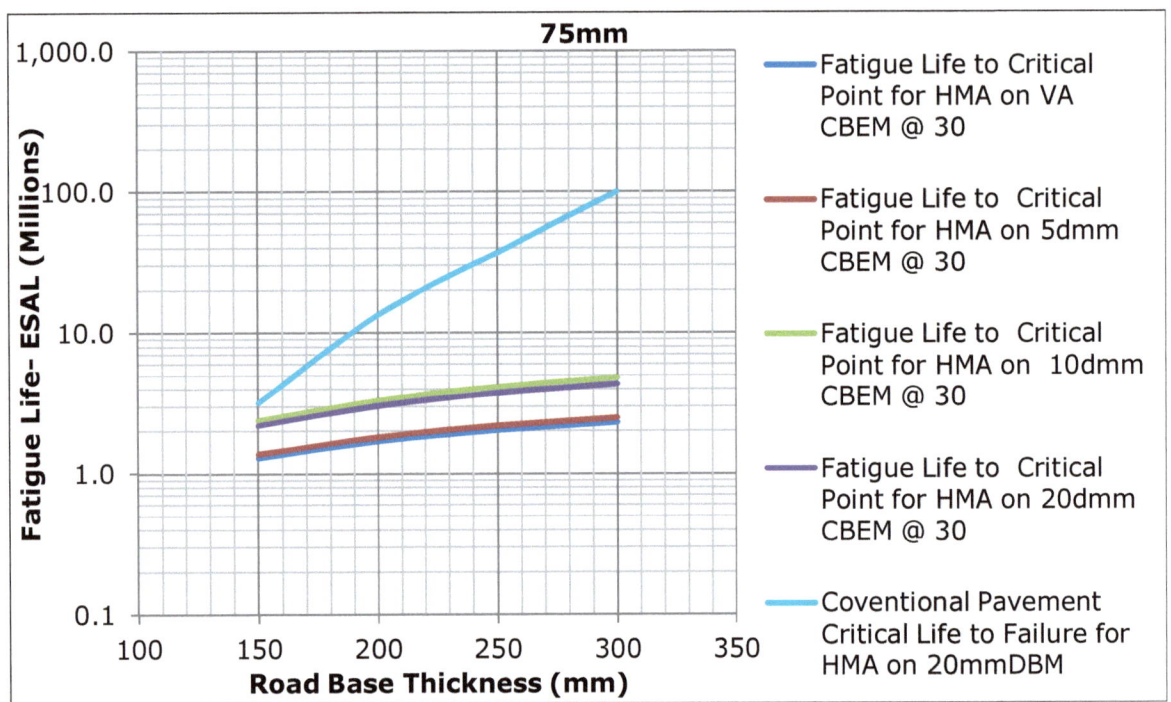

Figure 23: Effect of Road Base Layer Thickness and Material Type on Fatigue Lives to Critical Condition for Surfacing Thickness of 75mm HMA

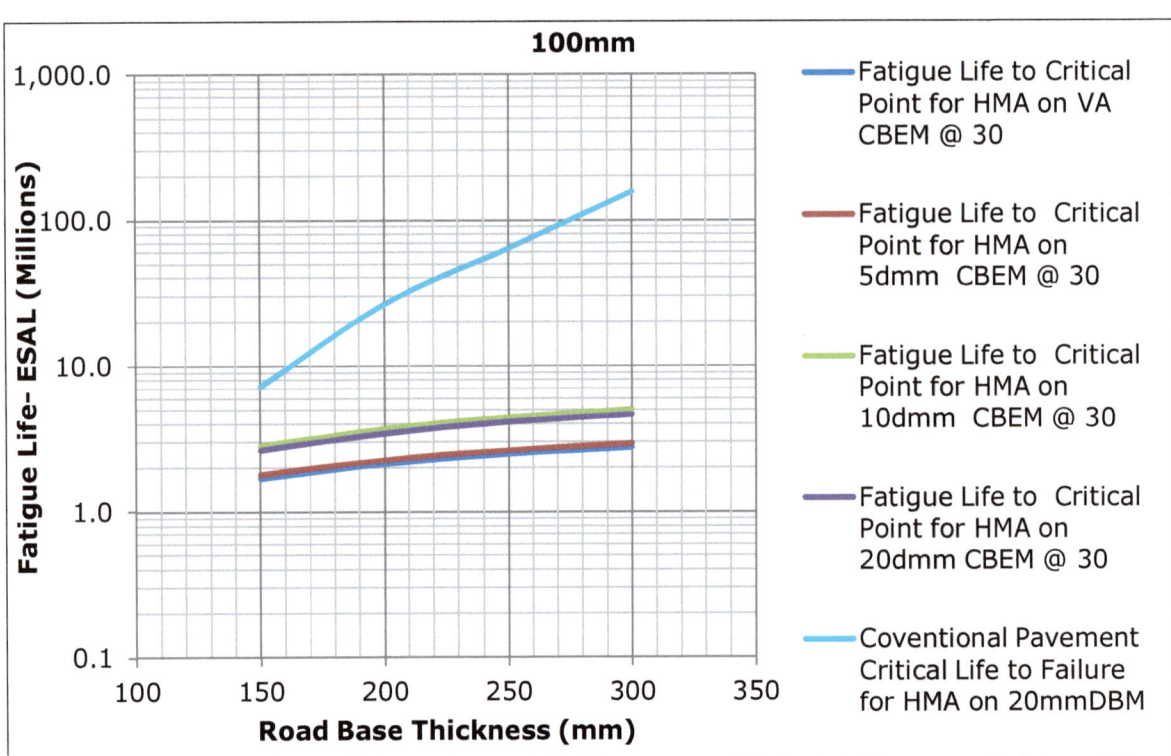

Figure 24: Effect of Road Base Layer Thickness and Material Type on Fatigue Lives to Critical Condition for Surfacing Thickness of 30mm HMA

suggests that thicknesses as low as 20mm HMA surfacings are achievable in pavements containing cold asphalt. It is worth noting that the trend for the conventional pavement containing HMA (28mm DBM) as road base had a positive trend for the 30mm HMA surfacing. The trends observed for the other cases i.e. 50mm, 75mm and 100mm surfacing layer thicknesses were positive i.e. as road base layer thickness increases, fatigue life (ESALs) increases irrespective of the material type used for the road base. The HMA 28mmDBM was significantly better than the CBEMs although for all the cases considered, the ESAL for HMA surfacings supported by the CBEMs were all clearly above 1 million ESALs for critical condition. This was particularly obvious for the 50mm HMA surfacing as the predicted lives were consistently higher than those for the 75mm and 100mm HMA surfacing layers. The modular ratio of the surfacing layer and the underlying CBEM layer might be partly responsible for these unexpected results. The pavements containing the CBEM layers as road base are inverted compared to the HMA in this regard i.e. modular ratio for CBEM pavements are $\geq$ 2, while modular ratio for the HMA pavement is < 1. Although the rutting lives of the subgrade have to be checked, the results suggest that it is not structurally and also economically viable to use surfacing thicknesses higher than 50mm in this regard. Asphalt Academy (2009) advised that the maximum HMA surfacing for pavements containing cold asphalt layers should be limited to 50mm.

Similarly, the observed fatigue lives of the 50mm HMA surfacing for critical condition which is between 2.8 and 10.5 million ESALs seem to be close to the observations of Liebenberg and Visser (2004) for similar pavements in which they recorded 30 million ESALs to failure. Overall, the RAP CBEMs performed better than the VACBEM with the 10dmmCBEM and 20dmmCBEM offering the best performances in ESALs.

9   The Effect of CBEM Road Base Layer Property on Subgrade Performance

The rutting potential of the subgrade under the various pavement structures analysed were also studied. The governing criterion here is the vertical strain on top of the subgrade. Equations 7.1 and 7.2 were used for the computations of ESALs as earlier described. Figures 25 to 27 are relevant here. The figures which are for life (ESALs) to critical condition, generally indicate that the HMA (28mm DBM) used for the road base also offers the best protection for the subgrade of all the materials studied. All the CBEMs showed similar responses i.e. positive trend such that ESALs increase with corresponding increase in road base

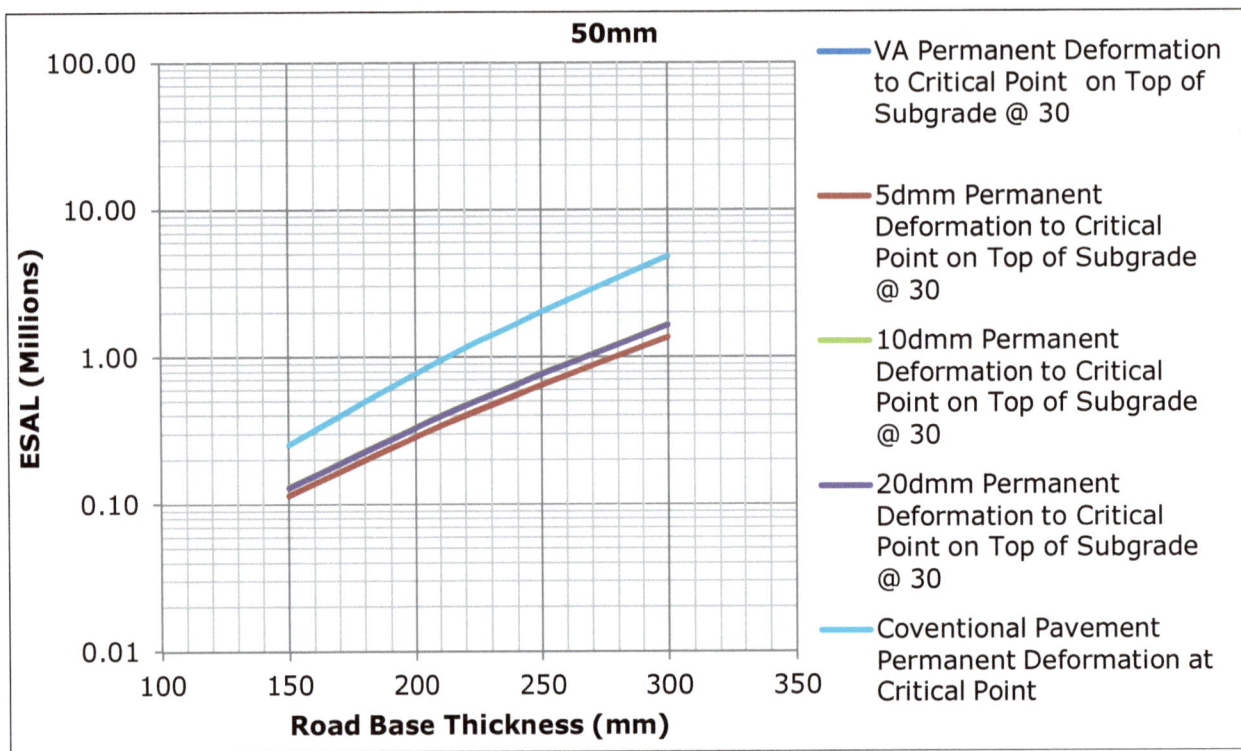

Figure 25: Effect of Road Base Layer Thickness and Material Type on Permanent Deformation to Critical Condition on top of Subgrade for Surfacing Thickness of 50mm HMA

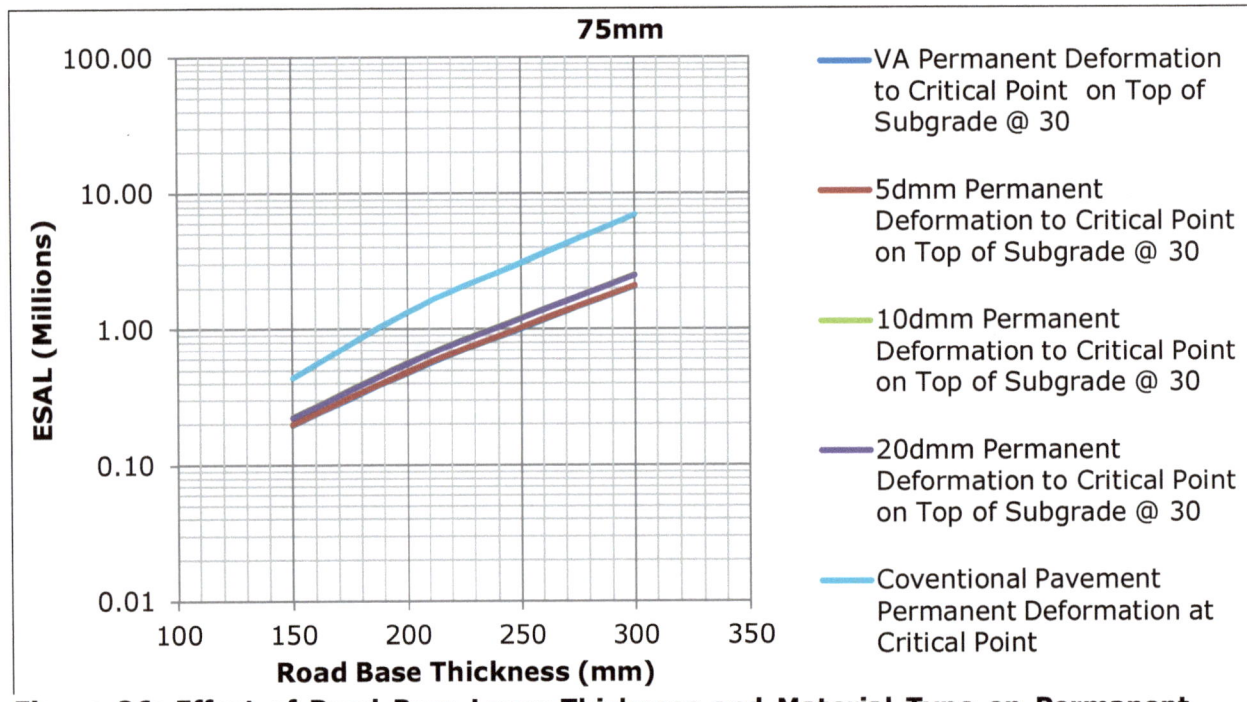

Figure 26: Effect of Road Base Layer Thickness and Material Type on Permanent Deformation to Critical Condition on top of Subgrade for Surfacing Thickness of 75mm HMA

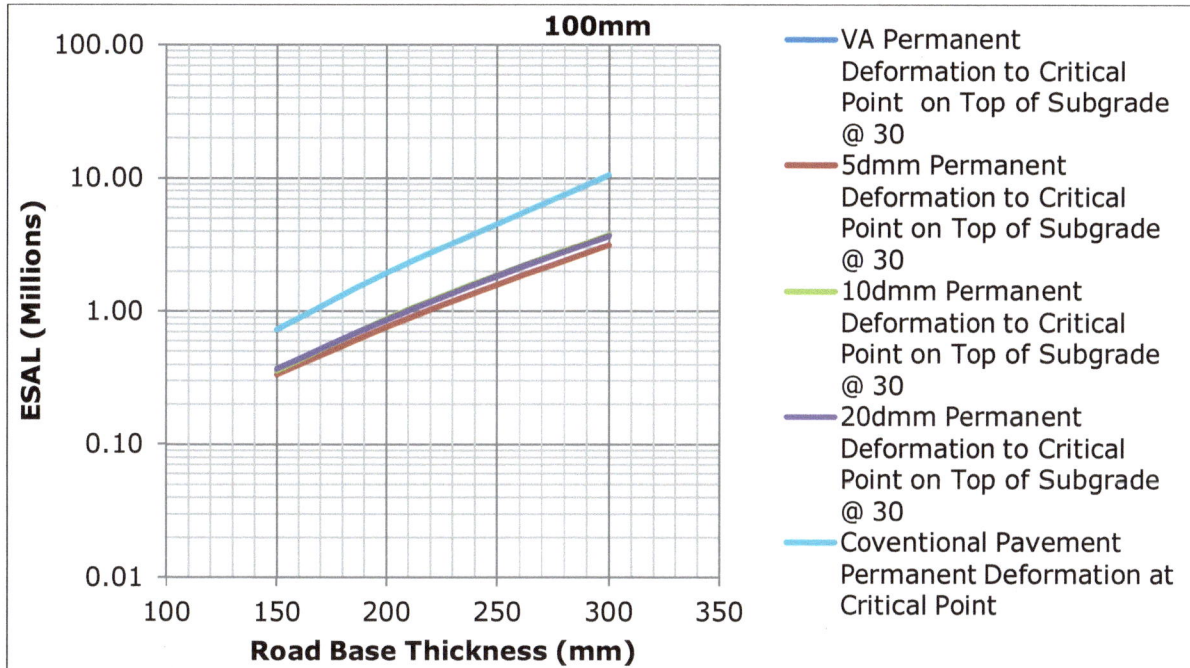

Figure 27: Effect of Road Base Layer Thickness and Material Type on Permanent Deformation to Critical Condition on top of Subgrade for Surfacing Thickness of 100mm HMA

thickness just as was the case for the conventional pavement. Both the 20dmmCBEM and 10dmmCBEM indicated the best protection for the subgrade of all the CBEMs studied.

The results showed that the surfacing layer had little or no significant influence on the response of the subgrade in the limited study carried out here whereas the road base thickness significantly influenced such responses. It is worth noting that the ESALs to critical condition for all the cases considered here were generally low compared to the ESALs obtained for the other parameters discussed earlier. The subgrade life also of course depends on the stiffness properties of the sub-base and subgrade. Although the design charts are hypothetical, they therefore suggest that the properties of the subgrade must be improved on in order to improve on the rutting responses of the subgrade and not necessarily the HMA surfacing and the CBEM road base layers.

10   Summary

In this book, Structural Design and Modelling of Flexible Pavements have been carried out. Although these examples are hypothetical in nature (sub-base and subgrade not varied in this work), the results are good indicators for ranking the road base materials. Along-side pavements containing CBEMs, conventional pavements were considered for comparative analysis. The following points can be drawn from the results obtained in this book:

- KENLAYER though time consuming, is a good analytical tool for analysis of flexible pavements involving non-linear elastic materials while BISAR 3.0 and OLCRACK are both good for linear elastic layers in multi-layer analysis of flexible pavements.
- Vertical stress distribution for pavements containing non-linear materials (CBEM) is similar irrespective of material type. Strains (vertical and horizontal) are however different.
- BISAR 3.0 has the limitation that it can only give similar results for stresses and strains as KENLAYER when the correct modulus values for each sub-layer considered are used.
- Except for ranking CBEMs, the use of maximum horizontal strains anywhere within the CBEM layers for design purposes is not sensitive to changes in thicknesses of the surfacing and CBEM layers. It is thus better to focus on checking the fatigue lives of surfacing and the rutting lives of subgrade for pavements containing CBEM layers.
- From the analysis done here and the subsequent design charts produced, the 28mm DBM and 5dmmCBEM indicated the best fatigue lives followed by 20dmmCBEM and 10dmmCBEM. VACBEM indicated the worst performance.
- Fatigue lives of the 50mm HMA surfacing for critical condition which is between 2.8 and 10.5 million ESALs suggest longer lives to failure although it is difficult to make predictions. Liebenberg and Visser (2004) recorded 30 million ESALs to failure for similar pavements.
- 28mm DBM gave very good support to 30/14 HRA surfacing layer compared to the CBEMs. A surfacing layer on HMA road base is highly sensitive to changes in thickness of the road base compared to CBEM road base. Such changes in HMA road bases resulted in significant increase in ESALs.
- The low stiffness values of the CBEMs did not affect their rutting resistance since road base made of either HMA or CBEM indicated similar responses in rutting resistance.
- The analysis showed that pavements containing HMA materials as surfacing and road base are better in performance compared to those made with CBEMs and thus confirms the findings/views of experts in the field.
- However, with all things being equal, pavements containing 5dmmCBEM/ 20dmmCBEM and probably 10dmmCBEM as road base should conveniently

accommodate about 5million ESALs before reaching the end of their service lives. The results presented here are generally for critical conditions, which imply failure would actually happen at larger ESALs values than presented in the design charts produced.

- The RAP CBEMs overall performed better than the VA CBEM in terms of ESALs applications that they offer.
- These RAP CBEMs would be applicable in low to medium volume traffic scenarios as road base materials.

11   Conclusion

The hypothetical examples presented in this book showed that KENLAYER, though time consuming, may be better than BISAR 3.0 for analysing pavements containing CBEMs as road base layers since the materials are non linear in elastic response. When KENLAYER seemed not to be sensitive enough in analysing the pavements with thin surfacing layer thicknesses of 30mm for horizontal strains at the bottom of the surfacing layer, OLCRACK proved in this regard, though the stiffness values calculated in the KENLAYER had to be used. However, in order for OLCRACK to have better application in modelling pavements containing non linear elastic layers, experimental modification reflecting the non linearity of the CBEMs would have to be introduced to the software. OLCRACK is basically good for estimating fatigue lives to failure for linear elastic layers in multi-layer analysis of flexible pavements.

The fundamental design criterion arrived at in this hypothetical design exercise was the horizontal strain at the bottom of the HMA surfacing layer for pavements containing CBEMs as road base layer. The less fundamental design criterion was the vertical strain on top of the subgrade. These two design criteria were chosen since, when using them, the type and thicknesses of CBEMs were reflected in the design ESALs. Though the result from the less fundamental criterion informs the use of improved subbase and subgrade, the RAP CBEMs offered better protection to the subgrade than the VACBEM. Similarly, the RAP CBEMs offered better support to the surfacing than the VACBEM.

Other design criteria that should have complemented those earlier stated are the fatigue and rutting lives of the CBEM layers. Common practice requires the use of the maximum strains in the layer. Although the CBEMs seemed to have been ranked, however, the use of such maximum strain values along with changes in the thicknesses of the layers were of no effect on the ESALs probably because of the modular ratio of the surfacing and CBEM layers, whereas it did for strain values at the bottom of the CBEM layers. It is not advisable to use the fatigue and rutting lives of the CBEM layers as design criteria for now. It is however possible that the representative strains required for considering CBEM layers in structural design of pavements can be ascertained through full scale pavement tests and this will be an interesting area for further research work.

References

Asphalt Academy (2009) *Technical Guideline: Bitumen Stabilised Materials, A Guideline for the Design and Construction of Bitumen Emulsion and Foamed Bitumen Stabilised Materials*. TG2, Second Edition, *Asphalt Academy*, Pretoria.

Brown, S. F. and Brunton, J. M. (1986) *An Introduction to Analytical Design of Bituminous Pavements*. Third Edition, University of Nottingham.

Brown, S.F. (1995) Practical test Procedures for Mechanical properties of Bituminous Materials. *Proc. Instn. Civ. Engrs Transp.*, 1998, 111, Nov, ICE, 289-297.

Ebels, L-J. (2008) *Characterisation of Material Properties and Behaviour of Cold Bituminous Mixtures for Road Pavements*. PhD Dissertation, Stellenbosch University.

Ekwulo, E. O. and Eme, D. B. (2009) Fatigue and Rutting Strain Analysis of Flexible Pavements Designed Using CBR Methods. *African Journal of Environmental Science and Technology*, Vol. 3 (12), Lagos.

Huang, Y. H. (2004) *Pavement Analysis and Design*. Pearson Prentice Hall, Upper Saddle River.

Jitareekul, P. (2009) *An Investigation into Cold In-Place Recycling of Asphalt Pavements*. PhD Thesis, University of Nottingham.

Liebenberg, J. J. E. and Visser, A. T. (2004) Towards a Mechanistic Structural Design Procedure for Emulsion-Treated Base Layers. *Journal of the South African Institution of Civil Engineering*, 46 (3), pg 2-8.

Oliveira, J. R. M. (2006) *Grouted Macadam: Material Characterisation for Pavement Design*. PhD Thesis, University of Nottingham.

Rahman, M. (2004) *Characterisation of Dry Process Crumb Rubber Modified Asphalt Mixtures*. PhD Thesis, School of Civil Engineering, University of Nottingham.

Read, J. M. (1996) *Fatigue Cracking of Bituminous Paving Mixtures*. PhD Thesis, University of Nottingham.University of Nottingham.

Read, J. M. and Whiteoak, D. (2003) *The Shell Bitumen Handbook*. Fifth Edition, Shell Bitumen, UK, Chertsey.

Shell Global Solutions (1998) *BISAR 3.0 Software and User's Manual*, Shell International Oil Products, The Hague.

Thom, N. H. (2000) A simplified Computer Model for Grid Reinforced Asphalt Overlays. *Proceedings of the 4th International RILEM Conference on Reflective Cracking in Pavements*, Ottawa, pg 37-46.

Thom, N. H. (2008) *Principles of Pavement Engineering*. Thomas Telford Publishing Limited, London.

Twagira, E. M. (2010) *Influence of Durability Properties on Performance of Bitumen Stabilized Materials*. PhD Dissertation, Stellenbosch University.

Yusof, M. A. (2005) *Investigating the Potential for Incorporating Tin Slag in Road Pavements.* PhD Thesis, University of Nottingham.